100道素菜

Vegetarian Meals

目錄
Contents

念念不忘素菜
UNFORGETTABLE VEGETARIAN DISHES

西式素菜
VEGETARIAN DISHES IN WESTERN STYLE

素食，吃出健康來！
Vegetarian Diet for Good Health!

近年標榜素食養生，多菜多豆多果仁，扭轉一般人葷食的概念。其實，有不少人對茹素仍存有疑惑～真的吃得飽嗎？攝取足夠鐵及鈣？會否貧血？

素食，可以吃得很豐盛。

芝麻合桃蓮藕餅、糖醋猴頭菇、素豆餅伴椰菜花飯、麻辣茄子豆腐、姬松茸節瓜雜豆湯……本書的主要材料除了以蔬菜為主，更重要的是補充豆類的營養，要知道豆類蘊含豐富的植物性蛋白質，而且可減少脂肪，提高免疫力，書內特別介紹用豆類做成素肉丸及肉扒，自行靈活配搭不同醬汁伴食，令吃素更多姿多采，肚子滿有飽足感！

想令素湯不再寡味，加添合桃、腰果是其中竅門，令湯水充滿肉香味。煲豆湯又如何？記緊放一塊陳皮，可下氣（降逆順氣）及增添豆湯香氣。

吃對了！身體好

肉類固然給我們帶來能量及人體必需礦物質，素食材料也富含鈣、鐵、鋅及其他維他命，為身體加注動力。提子乾、紅棗、蛋黃、海帶及黑豆可補充鐵質；堅果、豆腐、菠菜及芝士，是攝取鈣質的來源，進食天然的食材減低缺鐵缺鈣的情況，這是大自然給予的健康寶庫。

果仁也是健康的小吃，合桃、杏仁、腰果、開心果、松子仁等，壓碎後灑在餸菜上，香脆的口感令菜式生色不少。

你的身體響起警號？高膽固醇？肥胖？眼睛模糊？茹素可帶給你健康新體驗。

選對食油！事半功倍

本書所有食譜採用橄欖油、葡萄籽油或米糠油，橄欖油被公認為健康的食用油，令體內的低密度膽固醇（壞膽固醇）容易排出；但橄欖油只宜小火烹調，如炒煮建議使用葡萄籽油或米糠油，起煙點較高。葡萄籽油香味較淡，可配薑蒜等料頭炒煮，或伴醬汁食用增加香氣。

為了健康，今天開始素食行動！

There is a wave of vegetarian trend for living healthily, to have more vegetables, beans and nuts. However, people still have doubts towards a vegetarian diet.

Will I ever feel full?

Can I have enough iron and calcium? chances of anemia?

Vegetarian diet can be very rich and lavish.

Lotus root patties with walnut and sesame, deep-fried monkey-head mushrooms in sweet and sour sauce, cauliflower rice with bean patties, spicy eggplant and tofu, Agaricus Blazei mushroom, Chinese marrow and beans soup...... Vegetables are of course the main ingredients in this book, but more importantly it is the nutrition value of various beans. Beans are rich in plant based protein, they lower body fat and improve immunity. In this book I introduce vegetarian meat balls and burger made from beans, they can be matched with different sauces. They spice up your vegetarian diet and make your stomach full!

One trick of making soups is adding walnut and cashew nuts; they add a meaty flavour to your soups. How about bean soups? Put in dried tangerine peel to add aroma, it can also calm the Qi.

EAT THE RIGHT FOOD FOR GOOD HEALTH!

Meats of course can provide energy and essential minerals, but vegetarian ingredients have the same effect; they are also rich in calcium, iron, zinc and vitamins. Raisins, cranberries, egg yolk, kelp, black soybeans contain iron; nuts, tofu, spinach and cheese are sources of calcium. These ingredients contribute to our intake of iron and calcium, they are gifts from nature.

Nuts are healthy snacks, like walnut, almond, cashew nuts, pistachio, pine nuts etc. You can add a crunchy texture and flavours to your dishes with crushed nuts.

CHOOSE THE RIGHT OIL FOR MAXIMUM EFFECT!

The recipes in this book use olive oil, grapeseed oil and rice bran oil. Olive oil is widely accepted as a healthy choice. It helps expelling our low-density lipoprotein. However olive oil is only suitable for low heat cooking, it is recommended to use grape seed oil or rice bran oil for stir-fries as their smoke point is higher. Grapeseed oil has a relatively light taste, match with ginger, garlic or mix with sauce to add aroma and flavour.

For your health, let's go vegetarian today!

素食材料，你認識嗎？
Do you Know these Vegetarian Ingredients?

除了市售一般加工的素菜食材外，日常街市售賣的蔬菜或菇菌類，對素食人士都有很高的營養價值，你平時吃過了嗎？

Apart from the usual processed vegetarian food, have you had these nutritious, beneficial vegetables and mushrooms available in wet market?

蔬菜類
VEGETABLES

雅枝竹

降低血液內的膽固醇、三酸甘油脂，含高度抗氧化，有助改善脂肪肝、高血脂及尿酸等。食用時撕去外皮硬葉，取內芯食用，可作為冷盤或熱吃。

ARTICHOKE

It lowers cholesterol and triglyceride level in your blood vessels, alleviates fatty liver, high level of blood lipid and uric acid. It is an effective antioxidant. Tear off hard outer leaves and take only the inner part, it can be served cold or hot.

龍鬚菜

是合掌瓜的幼藤，含多種維他命、鐵、鈣及大量膳食纖維，熱量低，補充身體需要的葉酸及鐵質，有助消化、預防便秘及降血糖。涼拌或炒吃皆宜。

CHAYOTE SHOOTS

It contains various vitamins, iron, calcium. It is rich in dietary fibre and low in calories. It supplies folic acid and iron our bodies need. It alleviates digestion problems, constipation and high blood sugar. Suitable for cold or stir-fried dishes.

花生苗

從花生長出的幼苗，含豐富的蛋白質及膳食纖維，脂肪含量低，可增強抵抗力，降血脂及減重。炒煮或火鍋皆可，街市菜檔有售。

PEANUT SPROUTS

It is rich in protein and dietary fibre and low in fat. It strengthens our immunity, lowers blood lipid and body weight. Available in vegetable shops in wet markets, it is suitable for stir-fries or hotpot.

羽衣甘藍

近年被譽為的超級食品，蘊含多種抗氧化維他命，改善貧血、增強免疫功效、降血壓、整腸通便、抗老美肌、鞏固骨骼及牙齒等。羽衣甘藍常製成沙律，如吃不慣較硬的口感，可炒吃或包成餃子。

CURLY KALE

Seen as a superfood in recent years, it provides various anti-oxidising vitamins, strengthens immune system, lowers blood pressure, alleviate digestion and constipation, improves skin, bone and teeth conditions. It is usually used in salads; for a softer texture, stir-fried and made into dumplings.

枝豆

含豐富的食物纖維，改善便秘，降低膽固醇及血脂，其中所含的異黃酮類被稱為天然植物雌激素，消除令人體老化的活性氧，減緩機體老化。因枝豆纖維豐富，消化不良或腸胃不適者慎吃。

EDAMAME

Rich in dietary fibre, it alleviates constipation and high level of cholesterol and blood lipid. Its natural isoflavonoid is the equivalent of estrogen from plants, it can eliminate reactive oxygen species, thus slow down our body and organs from aging. Because of its rich fibre, people with digestive problem or stomachache should not eat it.

魚翅瓜

一絲絲的瓜肉含多種維他命、人體必須氨基酸及食用纖維，降血糖及膽固醇，熱量低，具清熱解毒功效，清甜味美，炒、燉及煲湯皆可。

FIG-LEAF GOURD

Its stringy flesh contains various vitamins, essential amino acids and dietary fibre. It is low in calories and lowers blood sugar and cholesterol. With its sweet and refreshing taste, it can also expel Heat and toxin-from our body. Stir fries, stews and soups can make use of it.

萵筍

爽口清脆，可涼拌或炒吃，是春季的當造蔬菜。其含鉀量高，促進排尿，有助高血壓及心臟病人士。由於可改善糖的代謝功能，血糖高者也適合食用。

CELTUCE

Spring is the season of celtuce. Its crunchy texture is suitable for cold or stir-fried dishes. It is rich in potassium, which promotes urination and benefits people with hypertension and heart disease. It also improves sugar/carbohydrate metabolism, which makes it suitable for people with high blood sugar.

綠色椰菜花

有豐富的葉綠素及葉黃素，維持視力健康，增強骨骼及牙齒生長，抗氧化高，有助降低癌症發生。選購時以花蕾緊密結實、葉片嫩綠及較重身的為佳，常見於每年10月至翌年4月。

GREEN CAULIFLOWER

It is rich in chlorophyll and lutein. It maintains our eyesight, promotes bone and tooth growth, and with its excellent anti-oxidation power, it helps lowering the possibility of cancer. Choose heavy cauliflowers with tight and close heads, and green leaves.They are usually available from October to April.

番薯苗

所含的抗氧化活性比其他蔬菜超出多倍，抗氧化功能強。而且含豐富的鎂，有助心血管的健康；維他命A可增強視力，預防感冒。用蒜頭炒吃或湯煮皆美味可口。

SWEET POTATO SHOOTS

Its anti-oxidation power is many times higher than other vegetables. It has rich magnesium, which benefits cardiovascular health; its vitamin A improves eyesight and prevent flu. Stir-fried with garlic or made into soup are two great ways to use them.

菇菌類
MUSHROOMS

姬松茸

是常用的保健菇菌，消炎、降血壓及血糖、抗敏感、抑制癌細胞、抵抗過濾性病毒等，提高糖尿病患者的免疫能力。乾的姬松茸呈橙黃色，具彈性及含豐富的香氣。

AGARICUS BLAZEI MUSHROOM

It is a popular mushroom for improving health. It alleviates inflammation, lowers blood pressure and sugar, suppresses cancer cells and viruses, improves immunity for people with diabetes. Dried ones are elastic, aromatic and has a yellow-orange colour.

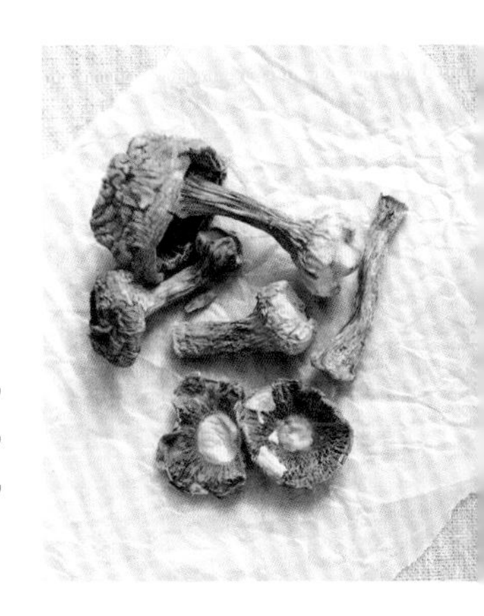

羊肚菌

菌傘如蜂巢而得名，健脾養胃，補腎補腦，預防感冒，以製成湯品最常見，市面有小顆及大顆的乾羊肚菌，功效相同。

MORCHELLA

Shaped like a honeycomb, it strengthens Spleen, Kidney and brain, tonifies Stomach and prevents flu. Usually made into soups, available in different sizes for the same effect.

猴頭菇

對消化不良、胃潰瘍、胃痛、胃脹及神經衰弱等有一定助益。猴頭菇香味充足；但帶鮮腥味，煮食前宜泡水及擠乾水分，可煲湯或炒吃，是素食者喜愛的食材之一。

MONKEY-HEAD MUSHROOM

It is beneficial to digestive problems, stomach ulcer, stomachache, swelling and neurasthenia (weak nerves). It is full of aroma and taste; it should be soaked and squeeze before use to remove the unpleasant earthy taste. It is one of the favourites among vegetarians, can be made into soup or stir-fries.

黃耳

具有很好的保健作用，蛋白質及膠質豐富，對肺熱痰多、氣喘及高血壓有幫助，提高身體的代謝功能。黃耳需浸泡及煲煮至軟身才烹調，炒煮燜煲湯皆可。

YELLOW FUNGUS

It has great health effects and is rich in protein and collagen. It alleviates Lung-Heat and phlegm, asthma and hypertension, improves metabolism. Before use, it should be soaked and boiled soft, it can be used for stir-fried, braised dishes or soups.

海竹笙

是藻類食品，富含海藻膠原蛋白，低脂、低膽固醇，有助整腸及排便。海竹笙可吸收湯汁的精華增添味道，於售賣素食材料的店舖有售。

BULL KELP

It is kind of seaweed. It is rich in collagen and low in fat and cholesterol. It is beneficial to digestion and bowel health. Bull kelp is excellent for absorbing the flavour of any sauce. It is available in vegetarian goods stores.

乾品類
DRIED GOODS

黑蒜

由新鮮蒜頭發酵而成，含較高的微量元素，是血管的清道夫，有效預防血管壁沉積的膽固醇及血脂，對三高有一定幫助。黑蒜可生吃或伴飯進食，口感甜糯。

BLACK GARLIC

Black garlic is made by fermenting fresh garlic. It has a relatively high amount of micronutrients, thus helps clearing the cholesterol and lipid from blood vessels. Black garlic can be served raw with rice, it is sweet and chewy.

桃膠

由桃樹自然分泌的汁液，具豐富的植物膠原蛋白，有養顏護膚的功效。桃膠口感煙韌、無味，需要搭配煮成糖水食用。

PEACH GUM

Natural gum from peach trees, it is rich in plant based collagen, which promotes skin conditions. Peach gum is chewy and tasteless, it should be used in a sweet soup.

椰棗

營養價值很高，含人體需要的多種維他命、礦物質及天然糖分，低脂、低膽固醇，具滋潤作用，可代替蜜棗使用。椰棗果肉甜，可當成日常果品。

DATE PALM

With high nutrition value, it contains various essential vitamins, mineral and natural sugar, and low amount of fat and cholesterol. Can be used to replace candied dates. It is nourishing and sweet it can be treated as an everyday fruit.

鷹咀豆（雞心豆）

是一種高蛋白、低脂及低熱量的健康食材，含豐富不飽和脂肪酸，可預防心血管疾病、降血糖、促進體內脂肪分解及代謝。鷹咀豆可煲湯、做成沙律或壓茸使用。

CHICKPEA

It is a healthy ingredients with high level of protein and low level of fat and calories. Its unsaturated fat prevents heart diseases, lowers blood sugar level and improve metabolism for fat. It can be used in soup, salad or hummus.

小扁豆（蘭度豆）

含蛋白質、維他命B及葉酸，其鐵質含量是其他豆類的兩倍，常煮湯或做成配菜。深色小扁豆有抗氧化、減緩衰老及預防心臟病的作用。毋須浸水，沖洗後可直接煮吃。

LENTIL

Lentil contains protein, vitamin B and folic acid. It has two times more iron than other beans. Mostly used to make soups and appetisers. Lentils with deeper colour is an antioxidant with effects of slowing aging and preventing heart diseases. There is no need to soak them before use, just rinse and cook.

甜竹（包括腐竹、枝竹）

由黃豆製成的豆製品，鈣質及卵磷脂豐富，預防血管硬化，保護心臟。甜竹味較甜，平時較少使用，多用作煮齋。腐竹口感滑溜；枝竹煙靭可口。

SWEET TOFU STICKS (INCLUDING DRIED TOFU SKIN, DEEP-FRIED TOFU STICKS)

It is made from soybeans and rich in calcium and lecithin. It prevents heart arteries from hardening and protects heart. Sweet tofu sticks are less used, sweet in taste, usually used for vegetarian dishes. Dried tofu skin has smooth texture; deep-fried tofu sticks are chewy.

小米

口感綿滑，營養價值比白米高，維他命B_1含量是五穀之冠。健脾和胃、補益身體、補血健腦，由於小米不含麩質，不刺激腸道，易被消化。

MILLET

Soft and smooth in texture, its nutrition value is higher than white rice. It has the most vitamin B_1 among five grains. It strengthens Spleen, brain, blood, calms Stomach and tonifies body. It is gluten-free thus it is easy to digest and does not trigger intestines.

木敦果

泰國盛產的乾果草本茶，利尿排毒、改善腹瀉及消化不良。日常調成冷熱飲品，拌入蜂蜜或檸檬汁，濃淡隨意。

BAEL

Mainly from Thailand, bael tea made from dried bael fruit, promotes urination, expels toxin, improves diarrhea and digestive problems. It can be made to cold or hot drinks and mixed with honey, lemon juice as desired.

做好準備，輕鬆煮素
Prepare Well and Easily Make Vegetarian Dishes

烹調素菜使用的醬料及食材，除了在街外購買，也可自己動手製作，只需花三兩個步驟，基本的醬汁及材料可為你變出百樣素菜。

While readymade sauces and ingredients for vegetarian dishes are available, you can also make it yourself in 2-3 easy steps. These sauces and ingredients serve as the basis for a variety of vegetarian dishes.

欖油羅勒醬
Pesto Sauce

素菜：芝士石榴橄欖番茄乾拌沙律菜p.44、素豆餅伴椰菜花飯p.106、素肉扒漢堡包p.124、欖油羅勒醬雙色小扁豆p.168、香烤小甘筍南瓜件p.204
Used in: Pomegranate, Olive and Tomato Salad with Cheese (p.44), Cauliflower Rice with Bean Patties (p.106), Vegetarian Hamburger (p.124), Green and Orange Lentil in Pesto Sauce (p.168), Roasted Baby Carrot and Pumpkin in Pesto Sauce (p.204)

| 材料 |

初榨橄欖油半碗

意大利黑醋4湯匙

青檸汁1湯匙

羅勒4棵

松子仁2湯匙

黑椒碎1茶匙

海鹽半茶匙

| 做法 |

1. 只取羅勒葉片，洗淨，切碎。
2. 將全部材料放進攪拌機，打磨成幼滑的欖油羅勒醬即可。

煮素小貼士

- 每次可做兩份，置於乾淨及無水分的玻璃瓶，最後加入初榨橄欖油1湯匙，蓋上保鮮紙，密封冷藏，可保存一星期。

| INGREDIENTS |

1/2 bowl extra virgin olive oil
4 tbsp Balsamic vinegar
1 tbsp lime juice
4 sprigs basil
2 tbsp pine nuts
1 tsp finely chopped black pepper
1/2 tsp sea salt

| METHOD |

1. Only pick the basil leaves. Rinse and finely chopped.
2. Put all the ingredients into the blender. Make a smooth paste and ready to use.

TIPS

- You can make 2 servings and put in a clean and dry glass container. Add 1 tbsp of olive oil on top, close with a plastic wrap and the lid. It can be stored for 1 week.

牛油果醬
Guacamole

素菜：鷹咀豆蔬菜薄餅卷p.110
Used in: Hummus Tortilla with Vegetables (p.110)

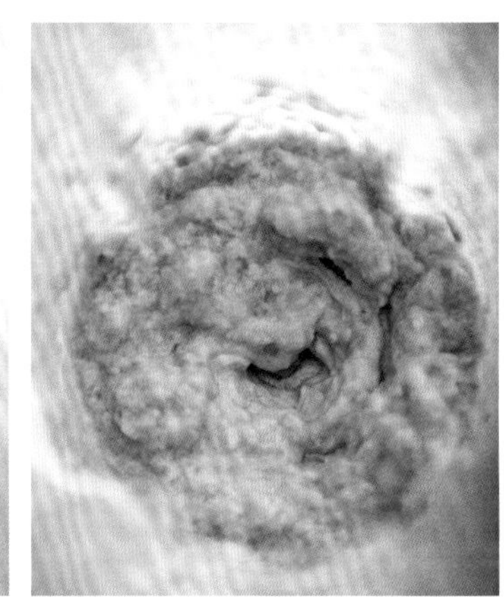

| 材料 |

熟牛油果2個（中型）

橄欖油4湯匙

檸檬汁2湯匙

黑椒碎1茶匙

海鹽1/3茶匙

| 做法 |

1. 牛油果去皮、去核，切細塊。
2. 將全部材料放進攪拌機，磨成幼滑的牛油果醬即可。

| INGREDIENTS |

2 medium mature avocado
4 tbsp olive oil
2 tbsp lemon juice
1 tsp finely chopped black pepper
1/3 tsp sea salt

| METHOD |

1. Skin and core avocado, cut into small pieces.
2. Blend all ingredients together in a blender into smooth guacamole.

煮素小貼士

・此醬伴吃青瓜、甘筍及紫椰菜等，滋味又有益。

TIPS

- Mixing guacamole with cucumber, carrot and red cabbage, it is a delicious salad and good for health.

素肉燥冬菇醬
Vegetarian Minced Pork Sauce

素菜：素肉燥冬菇醬乾扁四季豆p.82、素肉燥醬甜椒拌粉絲p.95、素肉燥冬菇醬番薯湯麵p.176、

Used in: Fried Snap Beans in Vegetarian Minced Pork Sauce (p.82), Vermicelli and Bell Pepper Mixed in Vegetarian Minced Pork Sauce (p.95), Sweet Potato Noodles in Vegetarian Minced Pork Sauce (p.176),

煮素小貼士

- 醬煮好後可分成小份，放入雪櫃急凍，每次食用時取出解凍、煮熱，可保存兩星期。

TIPS

- After the sauce is ready, it can be divided into many parts. It can be stored in freezer for 2 weeks; before use, defrost and heat up.

| 材料 |

乾冬菇4朵

杏鮑菇2個

薑3片（剁碎）

八角2粒

紹酒1湯匙

| 調味料 |

素蠔油2湯匙

老抽2茶匙

黃砂糖1茶匙

麻油2茶匙（後下）

| 獻汁 |

粟粉1茶匙

水2湯匙

| 做法 |

1. 冬菇去蒂，用水浸2小時，洗淨，切碎；杏鮑菇洗淨、切碎。
2. 燒熱鑊下油2湯匙，下薑茸炒香，加入冬菇、杏鮑菇炒勻，灒酒。
3. 加入八角、熱水1碗及調味料煮滾，轉細火煮10分鐘，下獻汁煮滾，拌入麻油，盛起備用。

| INGREDIENTS |

4 dried black mushrooms
2 oyster mushrooms
3 slices ginger (chopped)
2 star anises
1 tbsp Shaoxing wine

| SEASONING |

2 tbsp vegetarian oyster sauce
2 tsp dark soy sauce
1 tsp brown sugar
2 tsp sesame oil (added last)

| THICKENING GLAZE |

1 tsp cornflour
2 tbsp water

| METHOD |

1. Remove stalks from black mushrooms and soak for 2 hours, rinse and chop; rinse and chop oyster mushroom.
2. Heat wok and add 2 tbsp of oil, stir-fry ginger until fragrant. Add black mushrooms and oyster mushrooms and stir well. Mix in Shaoxing wine and stir-fry.
3. Add star anises, 1 bowl of hot water and seasoning and bring to boil. Turn to low heat and cook for 10 minutes. Add thickening glaze and bring to boil. Mix in sesame oil. Set aside for later use.

番茄醬
Tomato Sauce

素菜：蔬菜薯茸餅p.46、番茄燴素肉丸p.102、鷹咀豆蔬菜薄餅卷p.110、芝士菠菜意大利戒指雲吞p.115、素千層麵p.122
Used in: Fried Potato and Broccoli Patties (p.46), Vegetarian Meat Balls in Tomato Sauce (p.102), Hummus Tortilla with Vegetables (p.110), Tortellini in Tomato Sauce (p.115), Vegetarian Lasagna (p.122),

| 材料 |

熟紅番茄1斤4兩
洋葱半個
蒜肉3粒（切碎）
橄欖油2湯匙
黑椒碎1茶匙
海鹽半茶匙

| 做法 |

1. 番茄去蒂，洗淨，切碎；洋葱去皮，洗淨，切碎。
2. 平底鑊下油，放入洋葱、蒜肉炒香，加入番茄炒勻，下熱水1碗煮滾，轉細火煮15分鐘，待番茄呈糊狀，下調味料煮滾，盛起備用。

| INGREDIENTS |

750 g mature tomatoes
1/2 onion
3 cloves garlic (chopped)
2 tbsp olive oil
1 tsp finely chopped black pepper
1/2 tsp sea salt

| METHOD |

1. Remove stalks from tomatoes, rinse and chop; peel onion, rinse and chop.
2. Heat pan and add oil, stir-fry onion and garlic until fragrant. Add tomatoes and stir well. Add 1 bowl of hot water and bring to boil. Turn to low heat and cook for 15 minutes until the sauce thickens. To finish, add the seasoning and bring to boil. Set aside for later use.

煮素小貼士

- 最好先將洋葱或蒜茸等配料爆香，令番茄醬的味道更香氣四溢。

TIPS

- It is best to fry onion and garlic with oil first; it will strengthen the aroma and flavour of the tomato sauce.

大豆芽菜素上湯
Soybean Sprout Vegetarian Stock

素菜：西班牙蔬菜飯p.112、馬齒莧菜湯p.186、杞子枸杞菜湯p.202、
Used in:Vegetarian Paella (p.112), Purslane Soup(p.186), Qi Zi and Matrimony Vine Soup (p.202),

| 材料 |

大豆芽菜半斤（洗淨）
花生苗半斤（洗淨）
甘筍1條（約6兩）
薑2片

| INGREDIENTS |

300 g soybean sprouts (rinsed)
300 g peanut sprouts (rinsed)
1 carrot (about 225 g)
2 slices ginger

| 做法 |

1. 甘筍去皮，洗淨，切片。
2. 大豆芽菜用白鑊烘至半乾，放入湯煲，加入花生苗、甘筍、薑片、熱水9杯，大火煲滾，轉細火煲45分鐘，隔渣，留上湯備用。

| METHOD |

1. Peel carrot, rinse and slice.
2. Roast soybean sprouts in a plain wok until half-dried. Put soybean sprouts in a pot, add peanut sprouts, carrot, ginger and 9 cups of hot water. Bring to boil over high heat and turn to low heat and simmer for 45 minutes. Strain the soup and set the soup aside for later use.

煮素小貼士

- 大豆芽用白鑊烘乾，可去掉豆腥味。
- 用不完的素上湯，待冷後冷藏，約可保存3至4天。
- 焗齋的冬菇蒂剪下預留，加入可煲成素湯，香味更濃郁。

TIPS

- Roast soybean sprouts in a plain wok before use; it can expel the unpleasant grassy taste.
- Any leftover vegetarian stock can be stored in refrigerator for 3-4 days.
- Add the stalks from mushroom cooking the vegetarian stock, it smells fragrant and tastes good.

昆布素上湯
Kelp Stock

素菜：素肉燥冬菇醬番薯湯麵p.176、香菇昆布煮鮮枝竹p.184、昆布素上湯蕎麥麵p.218

Used in: Sweet Potato Noodles in Vegetarian Minced Pork Sauce (p.176), Simmered Tofu Stick with Black Mushroom and Kelp (p.184), Buckheat Noodles in Kelp Stock (p.218)

煮素小貼士

- 材料看似簡單，重點在於昆布的選擇，以及熄火加蓋待一會，令素上湯更香濃。

TIPS

- The ingredients look simple, but the trick is about choosing the right kind of kelp. After turning off the heat, remain the pot covered to further enhance the aroma and flavour.

| 材料 |

昆布1段（約20cm）

甘筍1條（約6兩）

薑2片

| 做法 |

1. 昆布剪成數段；甘筍去皮，洗淨，切片。
2. 將甘筍、薑片放入煲，加入清水8杯煲滾，轉細火煲半小時，加入昆布煲15分鐘，熄火焗15分鐘，隔渣（昆布留用），預留昆布上湯備用。

| INGREDIENTS |

1 section kelp (20cm long)
1 carrot (about 225 g)
2 slices ginger

| METHOD |

1. Cut kelp into several sections; peel carrot, rinse and slice.
2. Put carrot and ginger in a pot, add 8 cups of water and bring to boil. Turn to low heat and boil for 30 minutes. Add kelp and boil for 15 minutes. Turn off heat and remain covered for 15 minutes. Strain the stock and keep the kelp. Set the stock aside for later use.

冰豆腐
Frozen Tofu

素菜：五香醬毛豆煮冰豆腐p.88
Used in: Braised Frozen Tofu and Edamame (p.88)

| 做法 |

1. 將一塊厚身豆腐放入膠袋，排走空氣，紮緊袋口。
2. 放入冰格冷藏至冰硬即成。

| METHOD |

1. Put a cube of thick tofu in a plastic bag, press the air out of the bag and seal it.
2. Put in freezer until it becomes hard and frozen.

煮素小貼士

・選購石膏豆腐製作冰豆腐，效果較佳。

TIPS

- Use tofu made from gypsum for a better resulting frozen tofu.

全素
素扒
Vegetarian Burger

素菜：素肉扒漢堡包p.124
Used in: Vegetarian Hamburger (p.124)

| 材料 |

黃豆4兩
急凍青豆半碗
糙米飯1碗
冬菇4朵
胚芽半碗

| 調味料 |

黑椒碎1茶匙
海鹽半茶匙

煮素小貼士

- 素扒可冷藏保存3天；急凍保存10天，毋須解凍，以慢火烘熟即可。
- 如沒有糙米飯，可用白飯及粗麥皮拌勻。
- 青豆、黃豆加水磨成漿，取豆渣製成素肉扒，壓出來的豆漿不要浪費，加點水可煮成美味豆漿。

TIPS

- Vegetarian burgers can be stored up to 3 days in refrigerator; 10 days in freezer. They can be fried directly in a pan; there is no need to defrost.
- Cooked brown rice can be replaced with white rice or coarse oatmeals.
- Blend green peas, soybeans and water together, strain the pulps to make vegetarian burgers. Do not waste the leftover liquid, add water and boil to make soymilk.

| 做法 |

1. 黃豆洗淨，用水浸4小時，洗淨，放入煲內，加入適量水煲20分鐘，下青豆同煲10分鐘，隔去水分，過冷河，瀝乾水分。
2. 冬菇去蒂，洗淨，用水浸2小時，擠乾水分，剁碎備用。
3. 黃豆、青豆放入攪拌機，加入適量水磨成幼滑豆漿，放入煲魚袋壓去水分，豆渣留用。
4. 豆渣、糙米飯、冬菇、調味料拌勻，加入胚芽搓成粉糰，分成乒乓球大小，搓圓按扁成素扒，冷藏2小時。進食前放於平底鑊，用慢火烘至兩面金黃即成。

| INGREDIENTS |

150 g soybeans
1/2 bowl frozen green peas
1 bowl cooked brown rice
4 dried black mushrooms
1/2 bowl wheat germ

| SEASONING |

1 tsp finely chopped black pepper
1/2 tsp sea salt

| METHOD |

1. Rinse soybeans and soak for 4 hours. Rinse and transfer in a pot. Add water and boil for 20 minutes. Add green peas and boil for 10 minutes. Drain, rinse with cold water and drain.
2. Remove stalks from black mushrooms, rinse and soak for 2 hours. Squeeze until dry and finely chop.
3. Blend soybeans, green peas and water in a blender into smooth soymilk. Transfer the mixture in a muslin cloth, squeeze out the soymilk and keep the leftover soy pulps.
4. Mix soy pulps, brown rice, black mushrooms and seasoning together. Mix in wheat germ. Shape the mixture into balls, press into burger shape. Put the burgers in refrigerator. Before serving, fry the burger in a pan over low heat, until brown.

全素

素肉丸
Vegetarian Meat Balls

素菜：番茄燴素肉丸p.102
Used in: Vegetarian Meat Balls in Tomato Sauce (p.102)

煮素小貼士

- 做好的素肉丸冷藏可保存3天，不建議急凍。

TIPS

- Vegetarian meat balls can be stored up to 3 days in refrigerator. It is not recommended to freeze them.

| 材料 |

黃豆4兩
急凍青豆半碗
糙米飯1碗
冬菇4朵
胚芽半碗
已浸發木耳碎半碗
洋葱碎2湯匙

| 調味料 |

黑椒碎1茶匙
海鹽半茶匙
肉桂粉半茶匙

| 做法 |

與素扒的步驟相同，最後加入木耳及洋葱碎搓成粉糰，分成乒乓球大小，再搓成素肉丸冷藏，煮食時沾上適量乾粟粉即可。

| INGREDIENTS |

150 g soybeans
1/2 bowl frozen green peas
1 bowl cooked brown rice
4 dried black mushrooms
1/2 bowl wheat germ
1/2 bowl soaked black fungus
2 tbsp chopped onion

| SEASONING |

1 tsp finely chopped black pepper
1/2 tsp sea salt
1/2 tsp ground cinnamon

| METHOD |

Same method as vegetarian burger, mix in black fungus and chopped onion to the mixture. Shape into balls and put in refrigerator. Before serving, with cornflour.

素豆餅
Bean Patties

素菜：素豆餅伴椰菜花飯p.106、鮮蔬拌豆茸餅p.220
Used in: Cauliflower Rice with Bean Patties (p.106), Bean Patties with Vegetables (p.220),

全素

| 材料 |

眉豆4兩

青豆半碗

粟米半碗

麵包糠半碗

粗麥皮適量

| 調味料 |

黑椒碎1茶匙

海鹽半茶匙

| 做法 |

1. 眉豆洗淨，用水浸1小時，洗淨，放入煲內加入適量水煲20分鐘，下青豆同煲10分鐘，隔去水分，過冷河，瀝乾水分。
2. 眉豆、青豆、粟米分兩次放入攪拌機，加入適量水磨成幼滑豆漿，放入煲魚袋壓去水分，豆渣留用。
3. 豆渣加入麵包糠、調味料拌勻，搓成粉糰，分成乒乓球大小，搓圓，均勻地沾上粗麥皮，按扁成豆餅，冷藏備用。

| INGREDIENTS |

150 g black-eyed beans
1/2 bowl green peas
1/2 bowl corn kernals
1/2 bowl bread crumbs
coarse oatmeals

| SEASONING |

1 tsp finely chopped black pepper
1/2 tsp sea salt

| METHOD |

1. Rinse black-eyed beans, soak for 1 hour and rinse. Boil in a pot for 20 minutes. Add green peas and boil for 10 minutes. Drain, rinse with cold water and drain.
2. Put black-eyed beans, green peas and corn kernels in a blender in 2 batches, add water and blend into smooth mixture. Transfer the mixture in a muslin cloth, squeeze out any liquid and keep the leftover pulps.
3. Mix the pulps, bread crumbs and seasoning together. Shape the mixture into balls. Coat with coarse oatmeals evenly. Press into patties and refrigerate.

煮素小貼士 

- 可使用任何豆類，如蠶豆，重點是將豆類煲至軟腍。
- 素豆餅冷藏可保存3天，不建議急凍。

TIPS

- Any kind of beans can be used, like broad beans, as long as they are boiled soft.
- Bean patties can be stored up to 3 days in refrigerator. It is not recommended to freeze them.

涼拌小食
Cold Appetisers

藜麥杏仁車厘茄羽衣甘藍

全素

Curly Kale Salad with Cherry Tomato, Quinoa and Almond

| 材料 |

紅藜麥1/4量杯*
杏仁10粒
車厘茄6粒
羽衣甘藍3塊
紅洋葱絲少許
*電飯煲量米杯

| 汁料 | （調勻）

橄欖油2湯匙
黑醋半湯匙
黑胡椒、海鹽各少許

| 做法 |

1. 紅藜麥及水半杯放入煲，用小火煮15分鐘待水分乾，熄火焗5分鐘備用。
2. 羽衣甘藍加水浸15分鐘，沖洗乾淨，撕成細塊。
3. 車厘茄去蒂，洗淨；杏仁壓碎。
4. 羽衣甘藍加入汁料拌勻，上碟，鋪上適量藜麥、車厘茄及紅洋葱絲，淋上餘下汁料，灑上杏仁碎即成。

| INGREDIENTS |

1/4 measuring cup red quinoa*
10 almonds
6 cherry tomatoes
3 pieces curly kale
red onion shreds
*measuring cup from rice cooker

| DRESSING | (MIXED WELL)

2 tbsp olive oil
1/2 tbsp black vinegar
finely chopped black pepper
sea salt

| METHOD |

1. Put red quinoa and 1/2 cup of water in a pot. Cook over low heat for 15 minutes until dry. Turn off heat and remain covered for 5 minutes.
2. Soak curly kale for 15 minutes, rinse and tear into small pieces.
3. Remove stalks from cherry tomatoes, rinse; crush almond into pieces.
4. Mix curly kale and the dressing together and plate. Top with quinoa, cherry tomatoes and red onion, drizzle any remaining dressing and almond pieces. Serve.

煮素小貼士

- 藜麥及羽衣甘藍這兩種食材，都是超級食物，加上橄欖油及果仁，常吃對身體有益。

TIPS

- Quinoa and curly kale are both superfood. Together with olive oil and nuts, having them frequently are beneficial to your health.

麻醬拌杞子鮮淮山
Fresh Yam and Qi Zi in Sesame Sauce

全素

| 材料 |

新鮮淮山1小段（大條）

杞子2茶匙

| 汁料 |（調勻）

芝麻醬1.5湯匙

蘋果醋1湯匙

黃砂糖1茶匙

鹽半茶匙

| 做法 |

1. 鮮淮山削去外皮，洗淨，放入滾水飛水1分鐘，浸冰水至冷卻。
2. 杞子放入熱水沖洗，瀝乾水分。
3. 鮮淮山切成方粒，排上碟，加入杞子，淋上麻醬汁料供食。

| INGREDIENTS |

1 section large fresh yam
2 tsp Qi Zi

| DRESSING | (MIXED WELL)

1.5 tbsp sesame paste
1 tbsp apple vinegar
1 tsp brown sugar
1/2 tsp salt

| METHOD |

1. Peel fresh yam and rinse. Scald for 1 minute. Soak in ice water until cold.
2. Rinse Qi Zi in hot water, drain.
3. Dice yam and arrange on a plate. Add qi zi and the dressing. Serve.

煮素小貼士

- 淮山飛水後浸入冰水至涼，爽口好味。

TIPS

- Soaking yam in ice water makes it crunchy and more delicious.

合桃腰果拌菠菜
Spinach with Walnut and Cashew

全素

| 材料 |

菠菜苗8兩

合桃肉、腰果各1湯匙

| 調味料 |

麻油1/3湯匙

生抽1/4湯匙

| 做法 |

1. 菠菜苗原棵浸半小時，清洗乾淨。
2. 煮滾清水半鍋，放下菠菜灼2分鐘，撈起，過冷河，擠乾水分，切掉頭端，切段備用。
3. 合桃、腰果壓碎。
4. 菠菜苗排上碟，灑入調味料、合桃及腰果碎即可。

| INGREDIENTS |

300 g baby spinach
1 tbsp walnut
1 tbsp cashew nuts

| SEASONING |

1/3 tbsp sesame oil
1/4 tbsp light soy sauce

| METHOD |

1. Soak a whole sprig of spinach for 30 minutes and rinse.
2. Bring half pot of water to boil. Boil spinach for 2 minutes, remove and rinse with cold water. Squeeze until dry, cut off the root and cut into sections.
3. Crush walnut and cashew nuts.
4. Arrange spinach on a plate. Add the seasoning, walnut and cashew nuts. Serve.

煮素小貼士 

- 菠菜含葉酸及鐵質，可補血，令人面色紅潤；尤其缺鐵性貧血人士，可多進食菠菜；但菠菜含草酸，用滾水略灼可去除大部分草酸。

TIPS

- Spinach contains folic acid and iron, which help replenish our blood. Having spinach is recommended for people with iron deficiency anaemia. Boiling spinach removes most of the oxalic acid from spinach.

全素

藜麥帶子
Potato Quinoa Patties

| 材料 |

煮熟紅藜麥半碗

馬鈴薯1個（約5兩）

菠菜2兩

沙律菜少許

青檸汁2茶匙

| 調味料 |

黑椒碎半茶匙

海鹽半茶匙

| 做法 |

1. 馬鈴薯洗淨外皮，切成4塊，放碟上，隔水蒸20分鐘至全熟。
2. 馬鈴薯趁熱去皮，搓成薯茸，下調味料拌勻。
3. 菠菜洗淨，放入滾水灼1分鐘，撈起，擠乾水分，切碎。
4. 紅藜麥、菠菜加入薯茸拌勻，搓成藜麥球。
5. 平底鍋下橄欖油，放入藜麥球煎至一面金黃，翻轉另一面用鑊鏟輕輕按壓煎至金黃，盛起藜麥帶子，伴上沙律菜及青檸汁供食。

| INGREDIENTS |

1/2 bowl cooked red quinoa
1 potato (about 188 g)
75 g spinach
salad vegetables
2 tsp lime juice

| SEASONING |

1/2 tsp finely chopped black pepper
1/2 tsp sea salt

| METHOD |

1. Rinse potato and cut into quarters. Steam for 20 minutes, until fully cooked.
2. Peel the potato when it is hot, and mash well. Mix well with seasoning.
3. Rinse spinach and scald for a minute. Remove, squeeze until dry and chop.
4. Mix mashed potato with quinoa and spinach. Shape the dough into balls.
5. Heat pan and add olive oil. Fry the quinoa balls until one side browns, flip to another side and press lightly into patties, fry until brown. Plate the patties and serve with salad and lime juice.

煮素小貼士

・藜麥及菠菜混合薯茸後，要待涼才容易搓成形。

TIPS

- After mixing mash potato with quinoa and spinach, let it cool down; it will be easier to shape by hands.

全素

開心果洋葱拌凍番茄
Tomato Chunks Mixed with Pistachio and Onion

| 材料 |

紅洋葱1/6個

熟紅番茄2個

開心果仁10粒

| 汁料 |（調勻）

橄欖油2湯匙

青檸汁2茶匙

黃砂糖半茶匙

黑椒碎、海鹽各少許

| 做法 |

1. 番茄去蒂，洗淨，切塊；紅洋葱洗淨，切粒。
2. 開心果壓碎。
3. 番茄、洋葱加入汁料拌勻，放於雪櫃2小時，上碟，灑下開心果供食。

| INGREDIENTS |

1/6 red onion
2 mature tomatoes
10 pistachio nuts

| DRESSING | (MIXED WELL)

2 tbsp olive oil
2 tsp lime juice
1/2 tsp brown sugar
finely chopped black pepper
sea salt

| METHOD |

1. Remove stalks from tomatoes, rinse and cut into pieces; rinse red onion and chop.
2. Crush pistachio nuts.
3. Mix tomatoes and onion with the dressing. Refrigerate for 2 hours. Plate and top with pistachio nuts. Serve.

煮素小貼士

・如生吃番茄，要選購全熟的，美味又可口。

TIPS

- For serving tomatoes raw, choose fully mature ones for the best taste and texture.

全素

涼拌鮮竹海竹笙
Chilled Tofu Sticks and Bull Kelp

| 材料 |

急凍鮮腐竹1包
海竹笙約20條
炒香白芝麻1茶匙
芫茜少許

| 汁料 |（調勻）

麻油半湯匙
陳醋1湯匙
生抽1/3湯匙
黃砂糖1茶匙

| 做法 |

1. 海竹笙用水浸約1小時，放入滾水飛水1分鐘，撈起及過冷河，瀝乾水分。
2. 急凍鮮腐竹解凍，用凍開水浸一下，撈起，擠乾水分，切段。
3. 將鮮腐竹、海竹笙及汁料拌勻，置雪櫃待1小時，灑入白芝麻及芫茜即成。

| INGREDIENTS |

1 pack frozen tofu stick
20 strips bull kelp
1 tsp toasted sesame
coriander

| DRESSING | (MIXED WELL)

1/2 tbsp sesame oil
1 tbsp aged vinegar
1/3 tbsp light soy sauce
1 tsp brown sugar

| METHOD |

1. Soak bull kelp for 1 hour. Scald for a minute, rinse with cold water and drain.
2. Defrost tofu sticks, and soak in cold water. Squeeze until dry and cut into sections.
3. Mix tofu sticks, bull kelp and the dressing together. Refrigerate for 1 hour. Top with sesame and coriander. Serve.

煮素小貼士

- 海竹笙是海藻類食材，含有蛋白質、鈣、碘及礦物質等，爽口，可製成燜煮菜式或四季皆宜的涼拌菜，素食店有售。

TIPS

- Bull kelp is a kind of algae seaweed, it contains proteins, calcium, iodine and minerals. With its crunchy texture, it is suitable for stewing and chilled dishes in all seasons. It is available in vegetarian stores.

蛋奶素

煎南瓜水餃
Pan-fried Pumkin Dumplings

| 材料 |

日本南瓜肉8兩
芝士2片（撕成小塊）
乾葱茸1湯匙
餃子皮20塊

| 調味料 |

肉桂粉半茶匙
鹽半茶匙

做法

1. 日本南瓜洗淨，去皮，切大塊，上碟，隔水蒸20分鐘，趁熱搓成南瓜茸。
2. 乾葱茸放入油鑊炒香，盛起。
3. 南瓜茸、芝士、乾葱茸、調味料拌成餡料。
4. 每塊餃子皮放入1湯匙南瓜餡料，在皮邊沿塗上少許水，對摺按實皮邊。
5. 平底鑊排上南瓜餃子，加入熱水1/4杯，加蓋，用中火煮至水分收乾，加入1湯匙油煎至餃子皮微黃，盛起即可。

| INGREDIENTS |

300 g Japanese pumpkin
2 slices cheese (tear in pieces)
1 tbsp chopped shallot
20 pieces dumpling wrappers

| SEASONING |

1/2 tsp ground cinnamon
1/2 tsp salt

| METHOD |

1. Rinse pumpkin, cut off the skin and cut into large piece. Steam pumpkin on a plate for 20 minutes and mash pumpkin when hot.
2. Fry shallot with oil in a wok until fragrant, set aside.
3. Mix mashed pumpkin, cheese, shallot and seasoning together to make the dumpling filling.
4. Put 1 tbsp of the filling on each dumpling wrapper, dip water along the edge of the wrapper, fold and press the edge to make dumpling.
5. Arrange pumpkin dumplings on a pan and add 1/4 cup of water. Cover the lid and cook over medium heat until dried. Add 1 tbsp of oil and fry until slightly brown. Serve.

煮素小貼士

・選用日本南瓜製作，餃子的味道香糯好吃。

TIPS

- Use Japanese pumpkin to achieve a smooth texture and great fragrance and taste.

蛋奶素

芝士石榴橄欖番茄乾拌沙律菜

Pomegranate, Olive and Tomato Salad with Cheese

| 材料 |

沙律菜適量
紅石榴1湯匙
黑橄欖4粒
番茄乾1湯匙
車打芝士1片（切絲）

| 汁料 |

欖油羅勒醬3湯匙（見p.16）

| 做法 |

1. 沙律菜洗淨，瀝乾水分。
2. 沙律菜加入汁料拌勻，上碟，鋪上黑橄欖、番茄乾、紅石榴及車打芝士絲即成。

| INGREDIENTS |

salad vegetables
1 tbsp pomegranate seeds
4 black olives
1 tbsp dried tomatoes
1 slice cheddar cheese (shredded)

| DRESSING |

3 tbsp pesto sauce (see p.16)

| METHOD |

1. Rinse salad vegetables, drain.
2. Mix salad vegetables with the dressing and plate. Top with black olives, dried tomatoes, pomegranate seeds and cheddar cheese. Serve.

煮素小貼士

- 沙律菜拌上用羅勒製成的醬汁，香濃惹味，可隨意配上抗氧化及降血壓的紅石榴子，營養及美味兼備。
- 紅石榴除了製成沙律，也可榨成石榴汁，非常好味，但留意其果糖高，每次別飲用太多，建議每日喝半個石榴份量的果汁。

TIPS

- Pesto sauce made from basil mixed with salad vegetables, has strong flavour and aroma. Added with pomegranate seeds makes it even more nutritious and delicious.
- Apart from making salad, pomegranate seeds can be blended into juice. While it is very delicious, it contains high sugar. It is suggested to have half a pomegranate worth of juice per day.

蛋奶素

蔬菜薯茸餅
Fried Potato and Broccoli Patties

| 材料 |

西蘭花半斤
馬鈴薯2個（約10兩）
洋葱半個
忌廉3湯匙
牛油1片（約1湯匙）
沙律菜少許
檸檬汁1湯匙
番茄醬適量（見p.22）

| 調味料 |

黑胡椒1茶匙
海鹽半茶匙

| 做法 |

1. 西蘭花切細朵，用水浸半小時，洗淨。
2. 馬鈴薯洗淨外皮，放入煲加入適量水焓20分鐘，撈起；放入西蘭花焓5分鐘，撈起。
3. 洋葱去外衣，洗淨，切碎；西蘭花切碎。
4. 馬鈴薯去皮，搓成薯茸，加入調味料、忌廉、牛油拌勻，再下洋葱、西蘭花拌勻，搓成大粒丸子，用手輕按成薯餅，置雪櫃急凍至硬。
5. 進食前將蔬菜薯茸餅放入已預熱焗爐，用180° C上下火焗半小時，取出，配沙律菜、伴檸檬汁式及番茄醬供食。

| INGREDIENTS |

300 g broccoli
2 potatoes (about 375 g)
1/2 onion
3 tbsp whipping cream
1 slice butter (about 1 tbsp)
salad vegetables
1 tbsp lemon juice
tomato sauce (see p.22)

| SEASONING |

1 tsp finely chopped black pepper
1/2 tsp sea salt

| METHOD |

1. Cut broccoli into small pieces, soak for 30 minutes and rinse.
2. Rinse potatoes, boil in a pot for 20 minutes, remove; boil broccoli for 5 minutes, remove.
3. Peel onion, rinse and chop; chop broccoli.
4. Peel potatoes, mash and mix with seasoning, whipping cream and butter. Mix in onion and broccoli. Shape the mixture into balls, press flat into patties and freeze until hard.
5. Preheat the oven, bake potatoes patties at 180°C for 30 minutes. Remove and serve with lemon juice or tomato sauce drizzled salad vegetables.

煮素小貼士

・如覺薯餅有點膩口，可伴橙塊或橙汁進食。

・愛吃薯餅的朋友，不妨多做點急凍貯存，可放兩星期食用。

TIPS

- If the potato patties are too greasy for you, serve with orange slices or juice.
- The patties can be stored in freezer up to 2 weeks. If you like potato patties, you can make a lot and freeze them.

全素

涼拌青瓜片
Chilled Cucumber Slices in Sesame Sauce

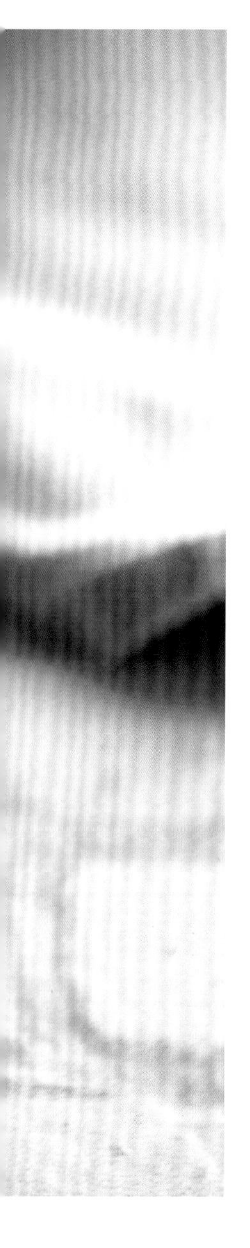

材料

中型青瓜1條
車厘茄6粒
炒香白芝麻1茶匙
海鹽3/4茶匙

汁料

麻醬1湯匙
陳醋1湯匙
麻油2茶匙
黃砂糖1茶匙

做法

1. 青瓜切掉頭尾兩端，洗淨，刨長片，灑入海鹽拌勻醃15分鐘，壓去汁液。
2. 車厘茄去蒂，洗淨，每粒切成2片。
3. 青瓜片上碟，伴上車厘茄，淋上汁料，灑入白芝麻食用。

| INGREDIENTS |

1 medium cucumber
6 cherry tomatoes
1 tsp toasted sesame
3/4 tsp sea salt

| DRESSING |

1 tbsp sesame paste
1 tbsp aged vinegar
2 tsp sesame oil
1 tsp brown sugar

| METHOD |

1. Cut off both ends of the cucumber, rinse and cut into long slices. Mix in sea salt and let it sit for 15 minutes. Squeeze out any excess water.
2. Remove stalks from cherry tomatoes, rinse and cut each one into 2 slices.
3. Arrange cucumber and cherry tomatoes on a plate. Top with the dressing and sesame. Serve.

煮素小貼士

・也可選用小青瓜製作，味雖淡，但更爽嫩，有不一樣的口感。

TIPS

- Small sized cucumbers can be used too. They have a less strong taste but a different crunchier texture.

蛋奶素

芝麻合桃蓮藕餅
Lotus Root Patties with Walnut and Sesame

材料

蓮藕1節（約6兩）
合桃肉8粒
白芝麻1湯匙
雞蛋1隻
粟粉1湯匙

調味料

胡椒粉少許
海鹽半茶匙

蘸汁

鎮江香醋1小碟

做法

1. 蓮藕洗淨污泥，削去外皮，刨成粗絲；合桃肉壓碎。
2. 蓮藕絲、合桃碎、白芝麻、雞蛋、粟粉、調味料全部拌勻。
3. 燒熱平底鑊下油3湯匙，舀起蓮藕漿1湯匙放入平底鑊，煎至兩面金黃香脆，上碟，伴蘸汁供食。

| INGREDIENTS |

1 section lotus root (about 225 g)
8 walnuts
1 tbsp sesame
1 egg
1 tbsp cornflour

| SEASONING |

pepper
1/2 tsp sea salt

| DIPPING SAUCE |

Zhenjiang vinegar

| METHOD |

1. Rinse off the the mud from lotus root, cut off the skin and shred into thick strips; crush walnuts.
2. Mix lotus root, walnut, sesame, egg, cornflour and seasoning together.
3. Heat pan and add 3 tbsp of oil. Add the lotus root mixture tablespoon by tablespoon, fry until both side browned. Plate and serve with the dipping sauce.

煮素小貼士

・將蓮藕刨成粗絲，咬入口啖啖蓮藕香，口感豐富。

TIPS

- Thick lotus root strips provide rich fragrance and texture to the patties.

全素

涼拌洋葱苦瓜片
Chilled Bitter Melon and Red Onion

| 材料 |

苦瓜1個（約8兩）

紅洋葱半個

| 汁料 |（調勻）

麻油2茶匙

蘋果醋1湯匙

青芥辣半茶匙

黃砂糖1茶匙

海鹽半茶匙

| 做法 |

1. 苦瓜開邊，刮淨籽及瓤，洗淨，切薄片。
2. 紅洋葱去外衣，切絲。
3. 將苦瓜片、洋葱絲放入冰水浸泡半小時。
4. 苦瓜片、洋葱絲及汁料拌勻，置雪櫃1小時可享用。

| INGREDIENTS |

1 bitter melon (about 300 g)
1/2 red onion

| DRESSING | (MIXED WELL)

2 tsp sesame oil
1 tbsp apple vinegar
1/2 tsp wasabi
1 tsp brown sugar
1/2 tsp sea salt

| METHOD |

1. Cut bitter melon length-wise in half. Remove all the seeds and core, rinse and cut into thin slices.
2. Peel and shred red onion.
3. Soak bitter melon and red onion in ice water for 30 minutes.
4. Mix bitter melon, red onion and the dressing together. Refrigerate for 1 hour. Serve.

煮素小貼士

・將苦瓜及洋葱泡於冰水，目的是令口感更爽脆。

TIPS

- Soaking bitter melon and red onion in ice water makes them crunchier.

滋潤素湯
Nourishing Vegetarian Soups

姬松茸節瓜雜豆湯

全素

Agaricus Blazei Mushroom, Chinese Marrow and Beans Soup

| 材料 |

姬松茸1兩
節瓜2個（約1斤）
雜豆半碗
椰棗4粒
陳皮1角

| 做法 |

1. 陳皮用水浸1小時，刮淨內瓤，洗淨。
2. 姬松茸用水浸1小時，洗淨。
3. 雜豆洗淨，用水浸1小時，隔去水分。
4. 節瓜刮淨外皮，洗淨，切大段。
5. 湯煲注入清水9碗，加入雜豆、節瓜、陳皮、椰棗，大火煲滾，轉小火煲1小時，加入姬松茸煲45分鐘，下少許海鹽調味即可。

| INGREDIENTS |

38 g Agaricus Blazei mushrooms
2 Chinese marrow (about 600 g)
1/2 bowl assorted beans
4 date palms
1 quarter dried tangerine peel

| METHOD |

1. Soak dried tangerine peel for 1 hour. Scrape off the pith and rinse.
2. Soak Agaricus Blazei mushrooms for 1 hour and rinse.
3. Rinse the beans, soak for 1 hour and drain.
4. Scrape off the peel of Chinese marrows, rinse and cut into large sections.
5. Put 9 bowls of water, beans, Chinese marrow, dried tangerine peel, date palms in a pot. Bring to boil over high heat. Turn to low heat and simmer for 1 hour. Add Agaricus Blazei mushrooms and boil for 45 minutes. Season with sea salt. Serve.

煮素小貼士

・姬松茸味道香濃，配上節瓜煲飲，祛濕消暑、健脾胃。

TIPS

- Agaricus Blazei mushrooms are rich in flavour and fragrance. When used with Chinese marrow in soups, it can expel the Heat and Dampness from your body, and strengthen the Spleen and Stomach.

全素

鮮淮山雪耳紅蘿蔔栗子湯

Fresh Yam, White Fungus, Carrot and Chestnut Soup

| 材料 |

鮮淮山半斤
紅蘿蔔1斤
雪耳1朵
栗子6兩
腰果2兩
陳皮1角

| 做法 |

1. 陳皮用水浸1小時，刮淨內瓤，洗淨。
2. 雪耳用水浸1小時，剪去硬蒂，撕成小塊，洗淨，飛水，瀝乾水分備用。
3. 鮮淮山洗去污泥，削皮，洗淨，切短度；紅蘿蔔去皮，洗淨，切大塊。
4. 栗子及腰果同洗淨。
5. 湯煲注入清水10碗，加入陳皮、紅蘿蔔、鮮淮山、栗子、腰果，大火煲滾，轉小火煲1小時，加入雪耳煲45分鐘，下少許海鹽調味即可。

| INGREDIENTS |

300 g fresh yam
600 g carrot
1 white fungus
225 g chestnuts
75 g cashew nuts
1 quarter dried tangerine peel

| METHOD |

1. Soak dried tangerine peel for 1 hour. Scrape off the pith and rinse.
2. Soak white fungus for 1 hour. Cut off the stems and cut into small pieces. Rinse and scald. Drain.
3. Rinse off the mud from fresh yam, peel and rinse. Cut into short sections. Peel carrot, rinse and cut into large pieces.
4. Rinse chestnuts and cashew nuts.
5. Put 10 bowls of water in a pot. Add dried tangerine peel, carrot, fresh yam, chestnuts and cashew nuts. Bring to boil over heat, turn to low heat and simmer for 1 hour. Add white fungus and boil for 45 minutes. Season with sea salt. Serve.

煮素小貼士

- 雪耳宜最後加入，以免煲煮太久而溶掉。
- 鮮淮山湯健脾補腎、滋潤肌膚，建議飲湯並吃湯料。

TIPS

- It is recommended to add white fungus last, to prevent it from dissolving.
- Fresh yams strengthen the Spleen, benefits the Kidney and nourish the skin. It is recommended to drink the soup and have all the ingredients.

全素

冬瓜冬菇竹笙芡實薏米湯
Winter Melon Soup with Black Mushroom, Zhu Sheng, Fox Nuts and Job's Tear

| 材料 |

冬瓜1.5斤
冬菇6朵
竹笙4條
芡實1兩
生薏米1兩
椰棗4粒
陳皮1角
薑3片

| 做法 |

1. 冬菇用水浸2小時，去蒂，洗淨。
2. 陳皮用水浸1小時，刮淨內瓤，洗淨。
3. 竹笙剪短度，與生薏米同浸半小時，洗淨。
4. 冬瓜去籽、去瓤，洗淨，切大塊。
5. 湯煲注入清水10碗，加入所有材料，大火煲滾，轉小火煲1.5小時，下少許海鹽調味即可。

| INGREDIENTS |

900 g winter melon
6 black mushrooms
4 pieces Zhu Sheng
38 g fox nuts
38 g Job's tears
4 date palms
1 quarter dried tangerine peel
3 slices ginger

| METHOD |

1. Soak black mushrooms for 2 hours, remove the stalks and rinse.
2. Soak dried tangerine peel for 1 hour. Scrape off the pith and rinse.
3. Cut Zhu Sheng into short section. Soak Zhu Sheng and Job's tears for 30 minutes. Rinse well.
4. Remove the seeds and core from winter melon, rinse and cut into large section.
5. Put 10 bowls of water in a pot. Add all the ingredients. Bring to boil over high heat. Turn to low heat and simmer for 1.5 hours. Season with salt. Serve.

煮素小貼士

・冬瓜湯消暑祛濕、清熱、健脾胃，可選用老黃瓜煲湯，功效相若。

TIPS

- Winter melon soup expels the Heat and Dampness, strengthen the Spleen and Stomach. Mature yellow cucumber can be used as well, with similar benefits.

全素

紅菜頭粟米甘筍腰果合桃湯

Beetroot Soup with Corn, Carrot, Cashew and Walnut

| 材料 |

紅菜頭1個（約8兩）

粟米2條

甘筍1條（6兩）

腰果、合桃各2兩

| 做法 |

1. 紅菜頭洗淨，削去外皮，切塊；甘筍洗淨，切塊；粟米去外衣，洗淨，切塊。
2. 腰果、合桃同洗淨。
3. 湯煲注入清水10杯，加入紅菜頭、甘筍、粟米、合桃、腰果，大火煲滾，轉小火煲1.5小時，下少許海鹽調味即可。

| INGREDIENTS |

1 beetroot (about 300 g)
2 ears sweet corn
1 carrot (225 g)
75 g cashew nuts
75 g walnuts

| METHOD |

1. Rinse beetroot, peel and cut into pieces; rinse carrot and cut into pieces; husk the corns, rinse and cut into pieces.
2. Rinse cashew nuts and walnuts.
3. Put 10 cups of water in a pot. Add beetroot, carrot, corn, walnut and cashew nuts. Bring to boil over high heat. Turn to low heat and simmer for 1.5 hours. Season with salt. Serve.

煮素小貼士

- 用紅菜頭煮成湯，清潤、補血，加入腰果或合桃煮成素湯，令湯水帶一陣肉香的味道。

TIPS

- Beetroot soup nourishes the body and energizes the blood. Adding cashew nuts and walnut to the soup introduces a hint of meat flavour.

全素

雙蓮冬菇花生紅綠豆湯
Lotus Root Soup with Lotus Seeds, Black Mushroom and Peanut

| 材料 |

蓮藕1斤
冬菇5朵
蓮子、花生、紅豆、綠豆各1兩
陳皮1角

| INGREDIENTS |

600 g lotus root
5 black mushrooms
38 g lotus seeds
38 g peanuts
38 g red beans
38 g mung beans
1 quarter dried tangerine peel

| 做法 |

1. 冬菇去蒂，用水浸2小時，洗淨，隔去水分。
2. 陳皮用水浸1小時，刮淨內瓤，洗淨。
3. 蓮子、花生、紅豆、綠豆同洗淨，用水浸半小時，隔去水分。
4. 蓮藕洗淨污泥，切厚塊。
5. 湯煲注入清水10碗，放入全部材料以大火煲滾，轉小火煲1小時45分鐘，下少許海鹽調味飲用。

| METHOD |

1. Remove stalks from black mushrooms and soak for 2 hours. Rinse and drain.
2. Soak dried tangerine peel for 1 hour. Scrape off the pith and rinse.
3. Rinse lotus seeds, peanuts, red beans and mung beans. Soak for 30 minutes and drain.
4. Rinse off the mud from lotus root and cut into thick slices.
5. Put 10 bowls of water in a pot. Add all ingredients and bring to boil over high heat. Turn to low heat and simmer for 1 hour 45 minutes. Season with sea salt. Serve.

煮素小貼士

・蓮藕湯補血補腎，令膚色更有光澤。

TIPS

- Lotus root energizes the blood and the Kidney and makes your skin brighter.

雙蓮冬菇花生紅綠豆湯

合掌瓜紅蘿蔔猴頭菇湯
Chayote Soup with Carrot and Monkey-head Mushroom

| 材料 |

合掌瓜1斤
紅蘿蔔12兩
乾猴頭菇4個
蓮子、百合各1兩
沙參、玉竹各半兩
椰棗4粒
陳皮1角

| 做法 |

1. 陳皮用水浸1小時，刮淨內瓤，洗淨。
2. 乾猴頭菇加水浸1小時，撕細塊，洗淨，飛水，過冷河，擠乾水分。
3. 蓮子去芯，用水浸1小時，隔去水分。
4. 百合、沙參、玉竹、椰棗同洗淨。
5. 合掌瓜切開邊，挖掉瓜核，洗淨，切大塊。
6. 紅蘿蔔削去外皮，洗淨，切塊。
7. 湯煲注入清水10碗，加入全部材料煲滾，轉小火煲1小時45分鐘，下少許海鹽調味。

| INGREDIENTS |

600 g chayote
450 g carrot
4 dried monkey-head mushrooms
38 g lotus seeds
38 g lily bulb
19 g Sha Shen
19 g Yu Zhu
4 date palms
1 quarter dried tangerine peel

| METHOD |

1. Soak dried tangerine peel for 1 hour. Scrape off the pith and rinse.
2. Soak dried monkey-head mushrooms for 1 hour. Tear into small pieces, rinse and scald. Rinse with cold water and squeeze until dry.
3. Remove the core from lotus seeds. Soak lotus seeds for 1 hour, drain.
4. Rinse lily bulb, Sha Shen, Yu Zhu and date palm.
5. Cut chayote in half, remove the seeds and core. Rinse and cut into large pieces.
6. Peel carrot, rinse and cut into pieces.
7. Put 10 bowls of water and add all ingredients. Bring to boil, turn to low heat and simmer for 1 hour 45 minutes. Season with sea salt. Serve.

煮素小貼士

・合掌瓜湯清潤、消暑、利濕。

・猴頭菇撕成小塊，有利釋出營養要素。

TIPS

- Chayote nourishes your body, expels the Heat and Dampness from your body.
- Tearing monkey-head mushrooms into small pieces, makes nutritions better dissolved in the soup.

全素

羊肚菌粟米紅蘿蔔淮山黑豆湯

Morchella Soup with Corn, Carrot, Yam and Black Soybean

| 材料 |

羊肚菌半兩

淮山1兩

黑豆2兩（用白鑊炒香）

粟米2條

紅蘿蔔1斤

椰棗4粒

陳皮1角

| 做法 |

1. 陳皮用水浸1小時，刮淨內瓤，洗淨。
2. 淮山用水浸1小時，隔去水分。
3. 羊肚菌用水浸半小時，剪開羊肚菌，洗淨內部砂粒，擠乾水分。
4. 粟米撕去外衣，洗淨，切塊；紅蘿蔔削去外皮，洗淨，切塊。
5. 湯煲注入清水10碗，加入所有材料（羊肚菌除外），大火煲滾，轉小火煲1小時，下羊肚菌煲半小時，下少許海鹽調味即成。

| INGREDIENTS |

19 g morchella

38 g dried yam

75 g black soybean (toasted in a plain wok)

2 ears sweet corn

600 g carrot

4 date palms

1 quarter dried tangerine peel

| METHOD |

1. Soak dried tangerine peel for 1 hour. Scrape off the pith and rinse.
2. Soak dried yam for 1 hour and drain.
3. Soak morchella for 30 minutes, cut open morchella and rinse off any dirt inside. Squeeze until dry.
4. Husk corns, rinse and cut into pieces; peel carrot, rinse and cut into pieces.
5. Put 10 bowls of water in a pot. Add all the ingredients (except morchella). Bring to boil over high heat. Turn to low heat and simmer for 1 hour. Add morchella and boil for 30 minutes. Season with salt. Serve.

煮素小貼士

· 羊肚菌及黑豆，具補腎、健脾胃之功效，可烏黑秀髮；黑豆用白鑊炒香再煲湯，效果更佳。

TIPS

- Morchella and black soybean benefit the Kidney, strengthen the Spleen and Stomach and darken hair. Toast black soybean in a plain wok before use for better effects

全素

番茄薯仔洋葱椰菜湯

Cabbage and Onion Soup with Tomato and Potato

| 材料 |

番茄3個
馬鈴薯1個
洋葱半個
西芹2棵
甘筍1個
椰菜半斤
蒜肉3粒

| 調味料 |

黑胡椒碎1茶匙
海鹽半茶匙

| 做法 |

1. 番茄去蒂，洗淨，切角；洋葱去外衣，洗淨，切碎。
2. 馬鈴薯及甘筍分別去皮，洗淨，切塊；西芹洗淨，切細塊；椰菜洗淨，切塊。
3. 湯煲下橄欖油2湯匙，加入洋葱、蒜肉炒香，加入番茄炒片刻，放入馬鈴薯、西芹、甘筍、椰菜，注入熱水9碗以大火煲15分鐘，轉小火煲1小時，下調味料拌勻即成。

| INGREDIENTS |

3 tomatoes
1 potato
1/2 onion
2 stalks celery
1 carrot
300 g cabbage
3 cloves garlic

| SEASONING |

1 tsp finely chopped black pepper
1/2 tsp sea salt

| METHOD |

1. Remove stalks from tomatoes, rinse and cut into wedges; peel onion, rinse and chop.
2. Peel potato and carrot, rinse and cut into piece; rinse celery and cut into small pieces; rinse cabbage and cut into pieces.
3. Add 2 tbsp of olive oil in a pot, stir- fry onion and garlic until fragrant, add tomato and fry. Add potato, celery, carrot, cabbage and 9 bowls of hot water. Bring to boil over high heat and boil for 15 minutes. Turn to low heat and simmer for 1 hour. Mix in seasoning. Serve.

煮素小貼士

- 湯煲好後添加1湯匙牛油，能提升整個湯的香味。
- 洋葱及番茄先炒一會，湯味更濃郁。

TIPS

- Adding 1 tbsp of butter before serving can enhance the aroma of the soup.
- Frying onion and tomato beforehand makes the soup tastes stronger.

全素

黃耳乾菌合桃魚翅瓜湯
Fig-leaf Gourd Soup with Yellow Fungus, Mushrooms and Walnut

| 材料 |

魚翅瓜1.5斤

黃耳半兩

姬松茸、竹笙、羊肚菌各半兩

合桃肉1.5兩

無花果3粒

椰棗2粒

薑1塊

| 做法 |

1. 黃耳用清水浸5至6小時，洗淨，切細朵。
2. 姬松茸、竹笙剪短度，洗淨；羊肚菌剪開，洗淨內部砂粒，擠乾水分。
3. 合桃肉、無花果、椰棗、薑塊同洗淨。
4. 魚翅瓜去籽，洗淨，切大塊。
5. 湯煲注入清水10碗煮滾，加入魚翅瓜，放入全部煲湯配料，大火煲15分鐘，轉小火煲1小時30分鐘，刮出魚翅瓜肉（皮不要），瓜肉放回煲內，下少許海鹽調味，煲滾即成。

| INGREDIENTS |

900 g fig-leaf gourd
19 g yellow fungus
19 g Agaricus Blazei mushroom
19 g Zhu Sheng
19 g morchella
56 g walnuts
3 dried figs
2 date palms
1 slice ginger

| METHOD |

1. Soak yellow fungus for 5 to 6 hours, rinse and cut into small pieces.
2. Cut Agaricus Blazei mushrooms and Zhu Sheng into short sections and rinse; cut open morchella and rinse off any dirt inside. Squeeze until dry.
3. Rinse walnuts, figs, date palms and ginger.
4. Remove seeds from fig-leaf gourd, rinse and cut into large pieces.
5. Bring 10 bowls of water in a pot to boil. Add fig-leaf gourd and all ingredients and boil over high heat for 15 minutes. Turn to low heat and simmer for 1 hour 30 minutes. Scrape the flesh of fig-leaf gourd and discard the skin. Put the flesh back in the pot. Season with sea salt. Bring to boil again and serve.

煮素小貼士

- 魚翅瓜湯具補腎、滋潤肌膚的功效，更可增添骨膠原，令關節及肌膚充滿彈性。
- 必須洗掉乾菌藏有的砂粒，徹底吸收菇菌的精華。

TIPS

- Fig-leaf gourd energizes the Kidney and nourishes the skin. With its collagen it enhance joints and skin.
- To achieve the best benefits, any dirt must be rinsed off from the mushrooms.

全素

南瓜粟米羹
Pumpkin and Corn Thick Soup

| 材料 |

日本南瓜12兩

粟米1條

水4碗

| 調味料 |

肉桂粉1茶匙

海鹽半茶匙

| 做法 |

1. 南瓜去籽，去皮，洗淨，切塊排上碟，隔水大火蒸15分鐘，備用。
2. 粟米去外衣，用刀削出粟粒。
3. 南瓜肉、水2.5碗放入攪拌機，磨成南瓜漿傾出；再加入粟米粒及餘下的水，磨成粟米漿。
4. 將南瓜漿、粟米漿放入煲內，用中小火煮滾（不可加蓋以免滾瀉），再煮10分鐘，一邊煮一邊攪拌，以免黏底，下調味料拌勻即成。

| INGREDIENTS |

450 g Japanese pumpkin

1 ear sweet corn

4 bowls water

| SEASONING |

1 tsp ground cinnamon

1/2 tsp sea salt

| METHOD |

1. Remove skin and seeds from pumpkin and rinse. Cut into pieces and arrange on a plate. Steam over high heat for 15 minutes. Set aside.
2. Husk corn, cut the corn kernels out.
3. Blend pumpkin and 2.5 bowls of water together in a blender, remove; blend corn kernels and the remaining water together.
4. Combine pumpkin and corn mixtures in a pot. Bring to boil over medium low heat, do not cover the pot to avoid boiling over. Stir constantly and cook for 10 minutes. Mix in seasoning. Serve.

煮素小貼士

- 可加入薑黃粉1茶匙調味，除了增添風味，對腦部也非常有益。
- 湯料容易黏底，緊記邊煮邊拌。

TIPS

- You can also add 1 tsp of turmeric powder to season; it adds flavour and benefits your brain.
- The soup mixture is easy to stick, remember to stir constantly.

蛋奶素

西蘭花青豆羹

Broccoli and Green Pea Thick Soup

|材料|

西蘭花12兩

急凍青豆3兩（解凍、洗淨）

洋葱半個

鮮忌廉3湯匙

麵粉2湯匙

牛油1湯匙

水4碗

|調味料|

黑椒碎半茶匙

海鹽半茶匙

| 做法 |

1. 西蘭花切細朵，用清水浸半小時，洗淨。
2. 煮滾半鍋水，放入西蘭花、青豆焓3分鐘，撈起，過冷河，瀝乾水分。
3. 用少量橄欖油炒香洋葱，備用。
4. 攪拌機內放入西蘭花、青豆、洋葱及水2.5碗，磨成西蘭花漿。
5. 牛油放鑊內，加入麵粉用小火炒勻，傾入餘下之水煮勻，轉入湯煲，加入西蘭花漿以慢火煮滾，邊煮邊攪拌，加入鮮忌廉煮5分鐘，下調味料煮滾即成。

| INGREDIENTS |

450 g broccoli
113 g frozen green peas (defrost and rinse)
1/2 onion
3 tbsp whipping cream
2 tbsp flour
1 tbsp butter
4 bowls water

| SEASONING |

1/2 tsp finely chopped black pepper
1/2 tsp sea salt

| METHOD |

1. Cut broccoli into small pieces, soak for 30 minutes and rinse.
2. Bring half pot of water to boil. Scald broccoli and green peas for 3 minutes, rinse with cold water and drain.
3. Stir-fry onion in olive oil until fragrant, set aside.
4. Blend broccoli, green peas, onion and 2.5 bowls of water together in a blender.
5. Put butter in a pan, mix in flour and fry over low heat. Add remaining water and mix well. Transfer the mixture in a pot, add broccoli mixture and bring to boil over low heat. Stir constantly while cooking. Add whipping cream and cook for 5 minutes. Add seasoning and bring to boil. Serve.

煮素小貼士

- 如全素者，可減去鮮忌廉，轉用椰奶煲煮；以及改用橄欖油炒麵粉糊，以代替牛油。
- 可烘香腰果加入攪拌，香氣四溢。
- 橄欖油可掩蓋青豆的草腥味，攪拌時可酌加。

TIPS

- If you are vegan, replace whipping cream with coconut milk; replace butter with olive oil.
- You can also add toasted walnut in the blender when making the broccoli mixture.
- Olive oil can cover the grassy taste of the green peas; you can add some to the broccoli mixture before blending.

念念不忘素菜
Unforgettable Vegetarian Dishes

糖醋猴頭菇
Deep-fried Monkey-head Mushrooms in Sweet and Sour Sauce

全素

煮素小貼士

- 乾猴頭菇一定要用水浸透，飛水，擠乾水分，炸好的猴頭菇才香酥，質感極像咕嚕肉。

TIPS

- Before using monkey-head mushrooms, they must be soaked thoroughly, scalded and squeezed dry; after deep-frying they are aromatic and puffy, have similar texture to deep-fried pork.

| 材料 |

乾猴頭菇7個
新鮮菠蘿1/4個
迷你三色甜椒各1隻
乾葱2粒（去衣、切片）
粟粉3湯匙

| 醃料 |

胡椒粉少量
生抽半湯匙
粟粉半茶匙

| 汁料 |（調勻）

茄汁3湯匙
青檸汁1湯匙
蘋果醋3湯匙
黃砂糖1湯匙
鹽半茶匙
粟粉2茶匙
水2湯匙

| 做法 |

1. 乾猴頭菇用水浸1小時，撕成小塊，飛水，過冷河，擠乾水分，下醃料拌勻醃味。
2. 菠蘿切細塊；三色甜椒去籽，洗淨，切粗粒。
3. 猴頭菇沾滿粟粉，放入滾油炸至金黃酥脆，撈起隔油。
4. 燒熱鑊下油1湯匙，下乾葱炒香，加入三色甜椒炒勻，傾下汁料煮滾至糊狀，加入菠蘿炒勻，再下猴頭菇拌勻即成。

| INGREDIENTS |

7 dried monkey-head mushrooms
1/4 fresh pineapple
1 mini red bell pepper
1 mini green bell pepper
1 mini yellow bell pepper
2 shallots (peeled, chopped)
3 tbsp cornflour

| MARINADE |

pepper
1/2 tbsp light soy sauce
1/2 tsp cornflour

| SWEET AND SOUR SAUCE | (MIXED WELL)

3 tbsp ketchup
1 tbsp lime juice
3 tbsp apple vinegar
1 tbsp brown sugar
1/2 tsp salt
2 tsp cornflour
2 tbsp water

| METHOD |

1. Soak dried monkey-head mushrooms for 1 hour. Tear into small pieces. Scald, rinse with cold water and squeeze until dry. Mix well with marinade.
2. Cut pineapple into small pieces; remove seeds from bell peppers, rinse and chop coarsely.
3. Coat monkey-head mushrooms with cornflour, deep-fry until golden brown. Drain.
4. Heat wok and add 1 tbsp of oil. Stir-fry shallot until fragrant. Add bell peppers and stir-fry. Add the sweet and sour sauce and cook until it thickens. Add pineapple and stir-fry well. Mix in monkey-head mushrooms. Serve.

蛋奶素

九層塔煎蛋
Fried Egg with Basil

| 材料 |

九層塔4棵

實板豆腐半塊

雞蛋3隻

薑茸1茶匙

| 調味料 |

胡椒粉少許

海鹽半茶匙

| 做法 |

1. 九層塔摘取嫩葉，洗淨，切碎。
2. 實板豆腐用清水略沖，壓爛。
3. 雞蛋拂匀拌散，加入豆腐、九層塔、薑茸、調味料拌匀。
4. 燒熱平底鑊下油2湯匙，傾下全部混合蛋漿，用中火煎至兩面金黃，盛起，切塊供食。

| INGREDIENTS |

4 stalks basil
1/2 cube firm tofu
3 eggs
1 tsp grated ginger

| SEASONING |

pepper
1/2 tsp sea salt

| METHOD |

1. Take the bail leaves, rinse and finely chop.
2. Rinse tofu briefly and crush.
3. Whisk the eggs and mix in tofu, basil, ginger and seasoning.
4. Heat pan and add 2 tbsp of oil. Fry the egg mixture over medium heat, until both sides browned. Remove and cut into pieces. Serve.

煮素小貼士

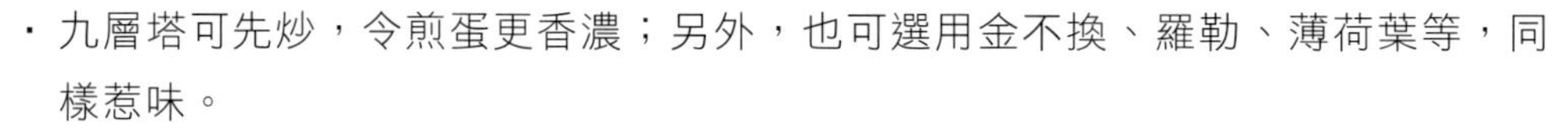

- 九層塔可先炒，令煎蛋更香濃；另外，也可選用金不換、羅勒、薄荷葉等，同樣惹味。

TIPS

- Basil can be fried first; it can enhance the flavour of the fried eggs; on the other hand, Thai basil, sweet basil, mint can be used too.

全素

七彩海竹笙素炒
Stir-fried Bull Kelp, Soybean Sprouts and Spiced Tofu

| 材料 |

海竹笙20段
迷你三色甜椒各1隻
大豆芽菜6兩
五香豆乾2塊
大頭菜1片
乾葱2粒（去衣、切片）

| 調味料 |

胡椒粉少許
黃砂糖半茶匙
素蠔油半湯匙

| 做法 |

1. 海竹笙用水浸1小時，放入滾水灼1分鐘，撈起，過冷河，隔乾水分。
2. 迷你三色甜椒去籽，洗淨，切粗絲；五香豆乾洗淨，切粗條；大頭菜洗淨，切粗絲。
3. 大豆芽菜洗淨，隔乾水分。
4. 燒熱鑊下油2湯匙，下乾葱炒香，加入大豆芽炒勻，下海竹笙、三色甜椒、五香豆乾、大頭菜炒片刻，下調味料炒勻，上碟供食。

| INGREDIENTS |

20 sections bull kelp
1 mini red bell pepper
1 mini green bell pepper
1 mini yellow bell pepper
225 g soybean sprouts
2 pieces spiced tofu
1 slice preserved white turnip
2 shallots (peeled, sliced)

| SEASONING |

pepper
1/2 tsp brown sugar
1/2 tbsp vegetarian oyster sauce

| METHOD |

1. Soak bull kelps for 1 hour. Scald for 1 minute, rinse with cold water and drain.
2. Remove seeds from bell peppers, rinse and shred roughly. Rinse spiced tofu and cut thickly. Presevered white turnip and shred roughly.
3. Rinse soybean sprouts and drain.
4. Heat wok and add 2 tbsp of oil, stir-fry shallots until fragrant. Add soybean sprouts and stir-fry well. Add bull kelps, bell peppers, spiced tofu, preserved white turnip and stir-fry. Mix in seasoning. Serve.

煮素小貼士

- 海竹笙是海藻類食材，必須用水浸泡至軟身才煮，可去除身體內的脂肪，常吃有益。

TIPS

- Bull kelp is a kind of algae seaweed, it must be soaked until soft before use. Bull kelp can expel excessive fat from the body, it is beneficial to have it frequently.

素肉燥冬菇醬乾扁四季豆

Fried Snap Beans in Vegetarian Minced Pork Sauce

| 材料 |

素肉燥冬菇醬1/3碗（見p.20）

四季豆8兩

蒜茸2茶匙

| 做法 |

1. 四季豆撕去老筋，洗淨，切短度。
2. 燒熱鑊下油2湯匙，下四季豆炒勻，加入蒜茸炒香，下半份素燥冬菇醬、熱水半碗煮片刻至水分收乾，上碟，鋪上餘下素肉燥冬菇醬食用。

| INGREDIENTS |

1/3 bowl vegetarian minced pork sauce (see p.20)
300 g snap beans
2 tsp grated garlic

| METHOD |

1. Tear off veins from snap beans, rinse and cut into short sections.
2. Heat wok and add 2 tbsp of oil, stir-fry snap beans well. Add garlic and stir-fry until fragrant. Add half of the vegetarian minced pork sauce and 1/2 bowl of water. Boil until the sauce dries up. Transfer on a plate. Add the remaining half of vegetarian minced pork sauce. Serve.

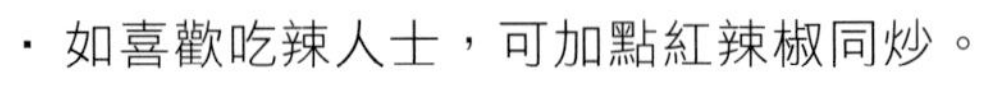

煮素小貼士

・如喜歡吃辣人士，可加點紅辣椒同炒。

TIPS

- For people who like spicy food, add red chilli and stir-fry together.

素肉燥冬菇醬乾扁四季豆

蛋奶素

煎釀小甜椒
Fried Bell Pepper Stuffed with Lotus Root and Peanut

材料

迷你三色甜椒6隻
蓮藕1小節
南乳肉花生2湯匙
雞蛋1隻（拂勻）
粟粉1湯匙

調味料

胡椒粉少許
海鹽半茶匙

做法

1. 蓮藕洗淨污泥，刨粗絲。
2. 南乳肉花生褪去外衣，壓碎。
3. 蓮藕絲、花生碎、蛋漿、調味料、粟粉拌勻成餡料。
4. 迷你三色甜椒切開、去籽，洗淨，抹乾水分，在內側塗上少許乾粟粉，釀入適量餡料。
5. 燒熱平底鑊，下油3湯匙，排入小甜椒（釀有餡料那面朝下）煎至金黃，翻轉另一面煎片刻，上碟供食。

INGREDIENTS

6 mini bell peppers (red, green, yellow)
1 section lotus root
2 tbsp tarocurd flavoured peanuts
1 egg (whisked)
1 tbsp cornflour

SEASONING

pepper
1/2 tsp sea salt

METHOD

1. Rinse off the mud from lotus root, shred coarsely.
2. Peel off peanuts skin and crush the peanuts.
3. Mix lotus root, peanuts, egg, seasoning, cornflour together to make the filling.
4. Cut open bell peppers and remove seeds. Rinse and wipe off any water. Coat the inside of bell peppers with cornflour and stuff the filling.
5. Heat pan and add 3 tbsp of oil. Fry the filling side of the bell peppers first until golden brown. Flip over and keep frying. Serve.

煮素小貼士

・必須抹乾甜椒內側的水分，並塗上乾粟粉，釀入的餡料才不脫落。

TIPS

- Any water must be wiped from the inside of bell peppers, and then coated with cornflour, to prevent the fillings falling off.

全素

南乳甜竹煮素菜
Vegetarian Stew with Sweet Tofu Stick and Fermented Tofu

| 材料 |

紹菜12兩
荷蘭豆2兩
甜竹2兩
炸枝竹2條
冬菇6朵
金針1兩
雲耳半兩
粉絲1.5兩
紅棗3粒（去核）
薑2片
大南乳半磚（搓爛）
黃砂糖1茶匙

| 做法 |

1. 冬菇去蒂，用水浸3小時，洗淨，切塊。
2. 金針剪去硬蒂，與雲耳一同用水浸1小時，飛水。
3. 炸枝竹折成短度，飛水，瀝乾水分。
4. 粉絲剪短度，用水浸半小時，瀝乾水分。
5. 荷蘭豆撕去硬筋，洗淨；紹菜切段，洗淨。
6. 燒熱鑊下油2湯匙，下薑片炒香，加入南乳、冬菇炒勻，注入熱水3碗，加蓋用中火煮15分鐘。
7. 加入金針、雲耳、紅棗、炸枝竹煮10分鐘，下甜竹、粉絲、紹菜、黃砂糖拌勻煮10分鐘，最後加入荷蘭豆拌勻煮3分鐘即可。

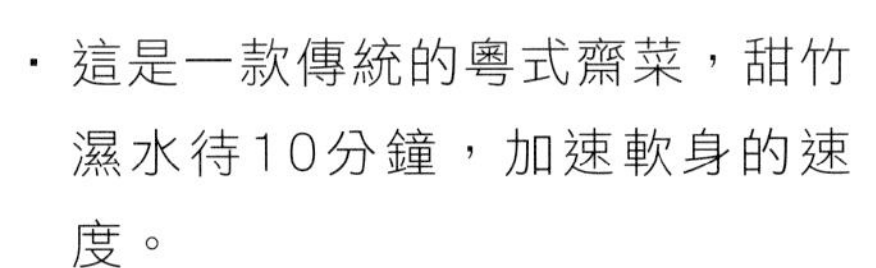

煮素小貼士

・這是一款傳統的粵式齋菜，甜竹濕水待10分鐘，加速軟身的速度。

TIPS

- This is a traditional Cantonese dish. Soak sweet tofu stick for 10 minutes before use; it will be softer.

| INGREDIENTS |

450 g Chinese cabbage
75 g snow peas
75 g sweet tofu stick
2 deep-fried tofu sticks
6 dried black mushrooms
38 g dried lily buds
19 g black fungus
57 g mung bean vermicelli
3 red dates (cored)
2 slices ginger
1/2 cube red fermented tofu (crushed)
1 tsp brown sugar

| METHOD |

1. Remove stalks from black mushrooms, soak for 3 hours. Rinse and cut into pieces.
2. Cut off hard stems from dried lily buds. Soak dried lily buds and black fungus together for 1 hour. Scald.
3. Tear deep-fried tofu stick into sections, scald and drain.
4. Cut mung bean vermicelli into sections. Soak for 30 minutes, drain.
5. Tear off hard veins from snow peas, rinse; cut Chinese cabbage into sections, rinse.
6. Heat wok and add 2 tbsp of oil, stir-fry ginger until fragrant. Add fermented tofu and black mushrooms and stir-fry. Add 3 bowls of water, cover the lid and cook over medium heat for 15 minutes.
7. Add dried lily buds, black fungus, red dates, deep-fried tofu stick and cook for 10 minutes. Add sweet tofu stick, mung bean vermicelli, Chinese cabbage and brown sugar. Mix well and cook for 10 minutes. To finish, add snow peas, mix well and cook for 3 minutes. Serve.

全素

五香醬毛豆煮冰豆腐

Braised Frozen Tofu and Edamame

| 材料 |

冰豆腐1磚（見p.25）
毛豆仁3兩
薑3片
麵豉醬1湯匙
八角2粒
麻油2茶匙

| 調味料 |

老抽1茶匙
黃砂糖1茶匙
水1碗

| 做法 |

1. 冰豆腐解凍，壓乾水分，切細塊。
2. 毛豆仁洗淨，放入滾水焓8分鐘，撈起備用。
3. 燒熱鑊下油2湯匙，下薑片拌香，加入麵豉醬拌勻，下調味料、八角煮滾，放入冰豆腐煮5分鐘，下毛豆仁煮5分鐘，灑入麻油拌勻，即可上碟。

| INGREDIENTS |

1 cube frozen tofu (see p.25)
113 g edamame
3 slices ginger
1 tbsp soybean paste
2 star anises
2 tsp sesame oil

| SEASONING |

1 tsp dark soy sauce
1 tsp brown sugar
1 bowl water

| METHOD |

1. Defrost frozen tofu, press to remove excessive water and cut into small pieces.
2. Rinse edamame, boil for 8 minutes, remove and set aside.
3. Heat wok and add 2 tbsp of oil, stir-fry ginger until fragrant. Add soybean paste and mix well. Add seasoning, star anises and bring to boil. Add frozen tofu and cook for 5 minutes. Add edamame and cook for 5 minutes. Mix in sesame oil. Serve.

煮素小貼士

- 自家製作冰豆腐非常簡單，只需用膠袋包紮豆腐，放入冰箱急凍24小時，煮食前2小時取出浸水解凍，緊記必須壓去水分。

TIPS

- It is easy to make frozen tofu: use a plastic bag to wrap a piece of tofu. Freeze for 24 hours. Before use, defrost for 2 hours and press to remove excessive water.

全素

欖菜蟲草花蒸豆泡勝瓜

Steamed Loofah with Preserved Olive and Cordycep Flowers

| 材料 |

勝瓜（絲瓜）12兩

豆泡2兩

乾蟲草花1兩

橄欖菜1湯匙

薑3片

| 調味料 |

胡椒粉少許

黃砂糖半茶匙

| 做法 |

1. 蟲草花用水浸半小時，洗淨，擠乾水分。
2. 豆泡切開，飛水，過冷河，擠乾水分。
3. 勝瓜削去外表硬皮，洗淨，開邊，切斜塊。
4. 豆泡排在蒸碟，鋪上勝瓜，隔水大火蒸5分鐘，保溫。
5. 燒熱鑊下油1湯匙，下薑片拌香，加入蟲草花、橄欖菜、調味料煮片刻，鋪在勝瓜上即成。

| INGREDIENTS |

450 g loofah

75 g deep-fried tofu

38 g dried cordycep flowers

1 tbsp preserved olive

3 slices ginger

| SEASONING |

pepper

1/2 tsp brown sugar

| METHOD |

1. Soak dried cordycep flowers for 30 minutes, rinse and squeeze until dry.
2. Cut open deep-fried tofu and scald. Rinse with cold water, squeeze until dry.
3. Cut off hard skin from loofah, rinse well, cut lengthwise in half and cut at an angle into pieces.
4. Arrange deep-fried tofu on a plate, arrange loofah on top. Steam over high heat for 5 minutes, keep warm.
5. Heat wok and add 1 tbsp of oil, stir-fry ginger until fragrant. Add cordycep flowers, preserved olive, seasoning and stir-fry. Put the mixture on top of loofah. Serve.

煮素小貼士

・橄欖菜非常惹味，但每個牌子鹹味不一，最好先試味再煮。

TIPS

- Preserved olive has a great strong taste, but each brand is different; it is recommended to taste it first before use.

全素

炸芋蝦
Deep-fried Taro Chips

| 材料 |

芋頭12兩
白芝麻2湯匙
糯米粉3湯匙

| 調味料 |

五香粉1茶匙
海鹽半茶匙

| 工具 |

罩籬2隻

| 做法 |

1. 芋頭削皮，洗淨，抹乾水分，用蘿蔔刨刨成粗條。
2. 芋絲加入調味料及白芝麻拌勻，再加入糯米粉拌勻。
3. 煮滾小半鍋油，罩籬放入熱油先浸10秒，放入1湯匙芋絲在罩籬，另一隻罩籬壓實芋絲，放入滾油炸片刻。
4. 取走上面的罩籬，退出芋蝦，回鍋炸至金黃，隔油，待冷食用（重複以上步驟至所有材料用完）。

煮素小貼士

・芋絲必須拌至乾身，如太濕可酌加糯米粉。
・新買回來的罩籬要放進熱油浸一會，以免芋絲黏着罩籬。

| INGREDIENTS |

450 g taro
2 tbsp sesame
3 tbsp glutinous rice flour

| SEASONING |

1 tsp five-spice powder
1/2 tsp sea salt

| TOOLS |

2 small strainer

| METHOD |

1. Cut off the skin from taro, rinse and wipe until dry. Shred into strips.
2. Mix taro with seasoning and sesame. Mix in glutinous rice flour.
3. Heat half pot of oil. Soak the first strainer in hot oil for 10 seconds. Add 1 tbsp of taro in the first strainer. Press the second strainer on the taro to shape the taro chip. Deep-fry the taro for a while.
4. Remove the second strainer and drop the taro chip. Deep-fry again until golden brown, let it cool and serve. Repeat the deep-frying process until all taro is used.

TIPS

- Taro strips must be mixed until dry. If the mixture is moist, add more glutinous rice flour.
- To prevent taro from sticking, new strainer must be soaked in hot oil for a while.

全素

素肉燥醬甜椒拌粉絲

Vermicelli and Bell Pepper Mixed in Vegetarian Minced Pork Sauce

材料

素肉燥冬菇醬1/3碗（見p.20）
粉絲1.5兩
甜椒半個（切粒）
腐皮2張
韭菜6條

汁料

素蠔油半湯匙
糖1茶匙
麻油2茶匙
胡椒粉少許
粟粉1茶匙
水3湯匙

做法

1. 粉絲剪短度，用水浸半小時，隔乾水分。
2. 素肉燥冬菇醬回鑊，加熱水3湯匙，下粉絲煮片刻，加入甜椒拌勻，盛起待冷備用。
3. 韭菜洗淨，放入滾水內灼軟，過冷河，備用。
4. 腐皮每塊剪成6大塊，用乾淨濕毛巾抹淨。
5. 腐皮2塊疊在一起，放入適量餡料，用韭菜紮成石榴果，隔水中火蒸5分鐘，取出。
6. 用小火煮滾汁料，淋在石榴果上即可。

煮素小貼士

・抹腐皮的濕毛巾不可太濕，以免腐皮糊化。
・餡料必須待涼才包入，以免濕氣影響腐皮，難以紮成石榴形狀。
・餡料可加入馬蹄，帶爽脆口感。

TIPS

- The moist towel should not be too wet, to prevent tofu sheet from dissolving.
- The filling must be cooled before wrapping, to prevent moisture affecting tofu sheets and shaping.
- Water chestnuts can be added in the filling to add a crunchy texture.

| INGREDIENTS |

1/3 bowl vegetarian minced pork sauce (see p.20)
57 g mung bean vermicelli
1/2 bell pepper (chopped)
2 tofu sheet
6 stalks chives

| SAUCE |

1/2 tbsp vegetarian oyster sauce
1 tsp sugar
2 tsp sesame oil
pepper
1 tsp cornflour
3 tbsp water

| METHOD |

1. Cut mung bean vermicelli into sections, soak for 30 minutes and drain.
2. Add vegetarian minced pork sauce and 3 tbsp of hot water. Add mung bean vermicelli and cook for a while. Mix in bell pepper to finish the filling. Remove and let it cool.
3. Rinse chives, scald until soft, rinse with cold water, set aside.
4. Cut each tofu sheet into 6 pieces. Wipe with a clean moist towel.
5. Put a piece of tofu sheet on top of another. Add filling, wrap into shape, tie a knot with chives. Steam over medium heat for 5 minutes.
6. Heat the sauce and pour on the dish. Serve.

全素

蠔汁三菇桃膠燴瓜甫
Braised Chinese Marrow with Mushrooms and Peach Gum

| 材料 |

鮮冬菇4朵
靈芝菇半包
杏鮑菇1小隻
桃膠1湯匙
節瓜2個（約1斤）
薑4片

| 調味料 |（調勻）

胡椒粉少許
麻油1茶匙
素蠔油1湯匙
粟粉1茶匙
蒸瓜汁3湯匙

| 做法 |

1. 桃膠用水浸4至5小時，挑去黑色污物，洗淨，放入滾水焓3分鐘，隔乾水分。
2. 節瓜刮淨外皮，直切，刮走瓜瓤，洗淨，切塊，排上蒸碟，灑入1/8茶匙海鹽，隔水中火蒸12分鐘，隔出汁液，保溫。
3. 鮮冬菇去蒂，洗淨，切粗條；杏鮑菇洗淨，切粗條；靈芝菇切去尾端，洗淨。
4. 燒熱鑊下油2湯匙，下薑片炒香，加入鮮菇、桃膠炒匀，下調味料煮3分鐘，淋上節瓜即成。

| INGREDIENTS |

4 fresh black mushrooms
1/2 pack marmoreal mushroom
1 small oyster mushroom
1 tbsp dried peach gum
2 Chinese marrows (about 600 g)
4 slices ginger

| SEASONING | (MIXED WELL)

pepper
1 tsp sesame oil
1 tbsp vegetarian oyster sauce
1 tsp cornflour
3 tbsp juice from steaming Chinese marrows

煮素小貼士

- 合掌瓜、青瓜、冬瓜等也非常合適燴煮，可吸收醬汁之精華。

TIPS

- Chayote, cucumber and winter melon are also suitable for this dish, as they can absorb the sauce well.

| METHOD |

1. Soak dried peach gum for 4-5 hours. Remove any black dirt. Rinse and scald for 3 minutes. Drain.
2. Peel off the skin from Chinese marrow, cut lengthwise in half. Rinse, cut into pieces and arrange on a plate. Add 1/8 tsp of sea salt and steam over medium heat for 12 minutes. Take the juice for the seasoning, keep the marrows warm.
3. Remove stalks from fresh black mushrooms, rinse and cut into thick strips; rinse oyster mushroom, cut into thick strips; cut off the root from marmoreal mushrooms and rinse.
4. Heat wok and add 2 tbsp of oil, stir-fry ginger until fragrant. Add all the mushrooms and peach gum and stir-fry well. Add seasoning and cook for 3 minutes. Transfer the mixture on top of Chinese marrows. Serve.

全素

素三杯雞
Stir-fried Tofu Skin Roll with Soy Sauce, Sesame Oil and Wine

材料

杏鮑菇2個

素雞2塊

九層塔2棵

乾葱（去衣、切片）

麻油2茶匙

調味料

老抽1湯匙

陳醋1湯匙

紹酒1湯匙

黃砂糖1茶匙

水3湯匙

做法

1. 杏鮑菇、素雞洗淨，切滾刀塊。
2. 九層塔摘取葉片，洗淨。
3. 燒熱鑊下油2湯匙，放入杏鮑菇、素雞煎至微黃，加入乾葱拌香，傾入調味料用中火煮5分鐘，待汁液收少，下麻油、九層塔拌勻，即可上碟。

INGREDIENTS

2 oyster mushrooms
2 pieces tofu skin roll (vegetarian chicken)
2 sprigs basil
shallot (peeled, sliced)
2 tsp sesame oil

SEASONING

1 tbsp dark soy sauce
1 tbsp aged vinegar
1 tbsp Shaoxing wine
1 tsp brown sugar
3 tbsp water

METHOD

1. Rinse oyster mushrooms and vegetarian chicken, cut both at an angle into pieces.
2. Take basil leaves and rinse.
3. Heat wok and add 2 tbsp of oil, stir-fry oyster mushrooms and vegetarian chicken until slightly browned. Stir-fry shallot until fragrant. Add seasoning and cook over medium heat for 5 minutes. Mix in sesame oil and basil. Serve.

煮素小貼士 

- 除了配搭素雞，也可煮炸豆腐或炸枝竹，同樣美味。

TIPS

- Apart from vegetarian chicken, deep-fried tofu or deep-fried tofu stick can be used too.

西式素菜
Vegetarian Dishes in Western Style

全素

番茄燴素肉丸
Vegetarian Meat Balls in Tomato Sauce

材料

素肉丸4個（見p.28）
番茄醬半碗（見p.22）
法包2片（烘脆）

做法

1. 素肉丸放入滾油，用中火炸至表面金黃，隔油。
2. 番茄醬煮滾，盛於深碟，放入素肉丸，伴已烘脆法包食用。

INGREDIENTS

4 vegetarian meat balls (see p.28)
1/2 bowl tomato sauce (see p.22)
2 slices baguette (toasted)

METHOD

1. Heat oil and deep fry vegetarian meat balls over medium heat until golden brown. Drain.
2. Heat the tomato sauce and transfer in a deep plate. Add the vegetarian meat balls and serve with baguette.

煮素小貼士 

- 素肉丸及素扒可一齊製作，先預備兩份基本食材，另備一份已處理的木耳及洋葱，可輕鬆做成不同口感的素肉。

TIPS

- Vegetarian meat balls and onion burger can be made together. Prepare 2 sets of ingredients and 1 set of black fungus and red onion, to make vegetarian meat with different textures.

蛋奶素

蘑菇番茄牛油果漢堡包
Hamburger with Avocado and Mushroom

| 材料 |

大蘑菇1個
大番茄1個
熟牛油果1個
紅洋葱圈少許
沙律菜適量
漢堡包2個
檸檬汁2茶匙

| 醬汁 |（拌勻）

芥末1茶匙
蜜糖2茶匙
純乳酪4湯匙

| 做法 |

1. 番加去蒂，洗淨，切厚片；大蘑菇用廚房紙抹淨，斜切厚片。
2. 牛油果去皮，去核，切厚片，用檸檬汁抹勻，以免氧化變黑。
3. 漢堡包放入焗爐，用150℃焗10分鐘，保溫。
4. 用橄欖油煎熱大蘑菇。
5. 漢堡包底層鋪好，排上沙律菜、番茄、蘑菇、牛油果、紅洋葱，淋上乳酪醬汁，蓋上漢堡包即可。

| INGREDIENTS |

1 large portobello mushroom
1 large tomato
1 mature avocadoe
red onion rings
salad vegetables
2 hamburger buns
2 tsp lemon juice

| SAUCE | (MIXED WELL)

1 tsp mustard
2 tsp honey
4 tbsp plain yoghurt

| METHOD |

1. Remote stalk from tomato, rinse and cut into thick slices; wipe portobello mushroom with kitchen pepper, cut at an angle into thick slices.
2. Cut off skin from avocado, core and cut into thick slices. Mix well with lemon juice to prevent oxidation.
3. Heat burger buns in an oven at 150°C for 10 minutes, keep them warm.
4. Fry portobello mushroom with olive oil until hot.
5. Place salad vegetables, tomato, portobello mushroom, avocado, red onion on top of the burger base bun. Top with the yoghurt sauce and finish with the burger top buns. Serve.

煮素小貼士

- 挑選牛油果有以下要訣：牛油果由綠色轉為黑色、果皮薄滑、手感有點軟的即代表熟。如牛油果呈全黑色、太腍即是壞透。

TIPS

- Avocados will turn brownish black from green as they mature. Choose soft avocados with black, smooth and thin skins. If they are overly black and soft, they are spoiled.

素豆餅伴椰菜花飯
Cauliflower Rice with Bean Patties

| 材料 |

素豆餅3塊（見p.29）
椰菜花8兩

| 醬汁 |

欖油羅勒醬2湯匙（見p.16）

| 做法 |

1. 椰菜花切成小粒狀，用水浸半小時，略沖洗。
2. 椰菜花放入滾水煮5分鐘，隔乾水分，加入醬汁拌勻上碟。
3. 平底鑊下橄欖油2湯匙，放入素豆餅煎至兩面金黃，排在椰菜花飯上即可。

| INGREDIENTS |

3 bean patties (see p.29)
300 g cauliflower

| SAUCE |

2 tbsp pesto sauce (see p.16)

| METHOD |

1. Dice cauliflower finely, soak for 30 minutes and rinse briefly.
2. Boil cauliflower for 5 minutes and drain. Mix cauliflower with the sauce and put on the plate.
3. Heat pan and add 2 tbsp of olive oil. Fry the bean patties until both sides browned. Arrange on top of cauliflower. Serve.

煮素小貼士

- 如想做成飯粒的效果，必須將椰菜花切成小顆粒狀。

TIPS

- To achieve the "rice" effect and texture, the cauliflower should be diced finely.

全素

番茄蘑菇羽衣甘藍意粉
Spaghetti with Curly Kale, Mushrooms and Tomato

| 材料 |

羽衣甘藍4塊
蘑菇4粒
番茄1個
蒜肉2粒（切片）
南瓜籽1湯匙
意粉80克
海鹽1茶匙

| 調味料 |

黑椒碎半茶匙

| 做法 |

1. 羽衣甘藍用水浸半小時，洗淨，撕成小塊。
2. 蘑菇用廚房紙抹淨，切細粒。
3. 番茄去蒂，洗淨，切細。
4. 煮滾半鍋水，放入鹽及意粉拌開，煮約8分鐘，撈起意粉（焓意粉水留用）。
5. 平底鑊放入橄欖油2湯匙，下蒜片拌香，加入番茄、蘑菇、羽衣甘藍拌勻，下意粉、焓意粉水3湯匙拌勻煮片刻，上碟，灑入南瓜籽食用。

| INGREDIENTS |

4 pieces curly kale
4 button mushrooms
1 tomato
2 cloves garlic (sliced)
1 tbsp pumpkin seeds
80 g spaghetti
1 tsp sea salt

| SEASONING |

1/2 tsp finely chopped black pepper

| METHOD |

1. Soak curly kale for 30 minutes, rinse and tear into small pieces.
2. Wipe mushrooms with kitchen paper, tear into small pieces.
3. Remove the stalk from tomato, cut into small pieces.
4. Boil half pot of water, add salt and cook spaghetti for about 8 minutes. Remove the spaghetti and keep the boiling water.
5. Heat 2 tbsp of olive oil. Stir-fry garlic until fragrant. Add tomato, mushrooms and curly kale and mix well. Add spaghetti and 3 tbsp of water from boiling spaghetti and cook. Put on the plate and top with pumpkin seeds. Serve.

煮素小貼士

・加入焓意粉水同煮，令意粉彈牙有嚼勁，更美味。

TIPS

- Cooking with the water from boiling spaghetti makes it more chewy and delicious.

全素

鷹咀豆蔬菜薄餅卷
Hummus Tortilla with Vegetables

| 材料 |

罐裝鷹咀豆半碗
小米半量杯（量米用杯）
青瓜絲適量
生菜絲適量
粟米薄餅皮2塊

| 醬汁 |

番茄醬（見p.22）
牛油果醬（見p.19）

| 拌豆汁 |

孜然粉半茶匙
黑椒粉半茶匙
檸檬汁半湯匙
橄欖油2湯匙

| INGREDIENTS |

1/2 bowl canned chickpeas
1/2 measuring cup millet (measuring cup from rice cooker)
shredded cucumber
shredded lettuce
2 pieces corn tortilla

| SAUCE |

tomato sauce (see p.22)
guacamole sauce (see p.19)

| SEASONING FOR HUMMUS |

1/2 tsp cumin
1/2 tbs ground black pepper
1/2 tbsp lemon juice
2 tbsp olive oil

煮素小貼士

・超市有已調味的瓶裝鷹咀豆醬，可省去拌豆汁的步驟。

TIPS

- Ready made hummus is also available in supermarkets, you can use those and omit the seasoning for hummus.

| 做法 |

1. 小米洗淨，用水1杯煮滾，轉中小火煮15分鐘，熄火焗10分鐘，盛起待冷備用。
2. 粟米薄餅皮放於平底鑊，用小火烘熱。
3. 鷹咀豆用叉壓成茸，加入小米、拌豆汁拌勻。
4. 粟米薄餅皮一塊鋪好，放入適量鷹咀豆、小米、生菜絲、青瓜絲，捲成筒型，以醬汁伴吃。

| METHOD |

1. Rinse millet, boil with 1 cup of water, turn to medium low heat for 15 minutes. Turn off heat and cover the lid for 10 minutes. Remove and let it cool.
2. Roast tortilla on a pan over low heat.
3. Mash chickpeas with a fork. Mix together with millet and the sauce for hummus.
4. Place hummus, lettuce, cucumber on the tortilla and roll into a cylinder. Serve with the sauces.

全素

西班牙蔬菜飯
Vegetarian Paella

|材料|（4人份量）

西班牙米250克
大豆芽菜素上湯3 3/4碗（見p.23）
大蘑菇2個
新鮮蠶豆6兩
四季豆4兩
椰菜花6兩
迷你紅、黃甜椒各1個
紅洋葱半個
番茄1個
蒜肉2粒（剁碎）

|調味料|

紅椒粉2茶匙
海鹽1茶匙

| 做法 |

1. 蠶豆用水煮滾，焓5分鐘，過冷河，去衣。
2. 椰菜花切細朵，用水浸半小時，洗淨。
3. 四季豆撕去老筋，洗淨，切段。
4. 迷你甜椒去蒂、去籽，洗淨，切粗條。
5. 大蘑菇用廚房紙抹淨，斜切厚片。
6. 紅洋葱去外衣，洗淨，切碎；番茄去蒂，洗淨，切碎。
7. 在西班牙飯盤下橄欖油3湯匙，加入洋葱、番茄、蒜茸炒香，下四季豆、椰菜花、甜椒拌勻，傾入大豆芽菜素上湯。(見圖1-2)
8. 加入西班牙米、蠶豆及調味料拌勻，用中小火煮約25分鐘，鋪上大蘑菇，待飯粒收乾水，熄火，蓋上廚房紙密封待15分鐘，即可食用。(見圖3-8)

*鳴謝：Félix Pérez González 老師提供西班牙飯的傳統家庭做法。

| INGREDIENTS | (4 SERVINGS)

250 g Spanish rice
3 3/4 bowls of soybean sprout vegetarian stock (see p.23)
2 portobello mushrooms
225 g fresh broad beans
150 g snap beans
225 g cauliflower
1 mini red bell pepper
1 mini yellow bell pepper
1/2 red onion
1 tomato
2 cloves garlic (chopped)

| SEASONING |

2 tsp paprika
1 tsp sea salt

| METHOD |

1. Boil broad beans for 5 minutes, rinse with cold water and peel.
2. Cut cauliflower into small pieces. Soak in water for 30 minutes and rinse.
3. Tear off hard veins from snap beans, rinse and cut into sections.
4. Remove stalks and seeds from bell peppers, rinse and cut into strips.
5. Wipe portobello mushrooms with kitchen paper, cut at an angle into thick slices.
6. Peel red onion, rinse and chop; remove stalk from tomato, rinse and chop.
7. Add 3 tbsp of olive oil in a paella pan. Stir-fry onion, tomato and garlic until fragrant. Add snap beans, cauliflower and bell peppers and mix well. Add soybean sprout vegetarian stock. (see pictures 1-2)
8. Add Spanish rice, broad beans and seasoning and mix well. Cook over medium low heat for about 25 minutes. Arrange portobello mushrooms on top. Cook until the rice dries up and turn off heat. Cover the paella with kitchen papers and let it sit for 15 minutes. Serve. (see pictures 3-8)

煮素小貼士

- 西班牙米毋須清洗，煮飯期間如水分不足，可酌加滾水；蔬菜可隨季節或個人喜好而轉換。
- 煮西班牙飯的重點是最後蓋上廚房紙待15分鐘，讓米粒均勻地熟透。
- 去皮後的蠶豆，令口感更綿滑。

TIPS

- There is no need to rinse Spanish rice. If the rice becomes too dry when cooking the paella, add boiling water. You can swap the vegetables depending on seasons and your taste.
- The trick of making paella is to cover the pan with kitchen papers for 15 minutes. The rice will be cooked evenly and thoroughly.
- Broad beans' texture is smooth and soft after peeling.

蛋奶素

芝士菠菜意大利戒指雲吞
Tortellini in Tomato Sauce

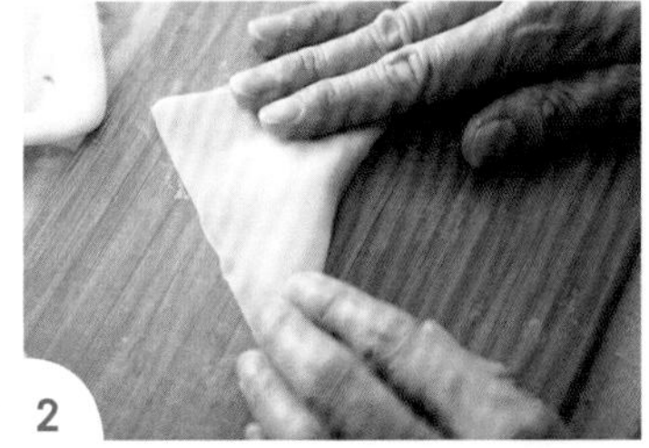

| 材料 |（可製成兩份）

菠菜苗6兩

車打芝士3片（切絲）

上海雲吞皮20塊

番茄醬1碗（見p.22）

| 調味料 |

黑胡椒碎1茶匙

海鹽半茶匙

羅勒1茶匙

| 做法 |

1. 菠菜苗用水浸半小時，切掉鬚根，洗淨，放入滾水灼1分鐘，撈起，過冷河，擠乾水分，切碎備用。
2. 將菠菜苗、車打芝士、調味料拌勻成餡料。
3. 取雲吞皮1塊，放入約1茶匙餡料，在雲吞皮邊沿掃上少許水，對摺成三角形，沿皮邊按實，排走空氣，捲成長條，用水按緊，將頭尾兩端按實成指環形狀（見圖1-5）。
4. 煮滾清水半鍋，放入芝士菠菜雲吞，用中火焓5分鐘即熟，撈起。
5. 煮滾番茄醬，加入芝士菠菜雲吞煮1分鐘即可。

煮素小貼士

・建議餡料不要包入太多，必須壓走空氣，捲得緊實即可。

| INGREDIENTS |

225 g young spinach
3 slices cheddar cheese (shredded)
20 pieces Shanghainese dumpling wrappers
1 bowl tomato sauce (see p.22)

| SEASONING |

1 tsp finely chopped black pepper
1/2 tsp sea salt
1 tsp basil

| METHOD |

1. Soak spinach for 30 minutes, cut off the roots and rinse. Boil for 1 minute and rinse with cold water. Squeeze until dry and finely chop.
2. To make the filling, mix spinach, cheddar cheese and seasoning together.
3. For each sheet of dumpling wrapper, put 1 tsp of filling, spread water along the edge of the wrapper. Fold into a triangle, press the edges together and press out any air inside. Roll up the dumpling and join the two ends together with water, shape into a ring. (see pictures 1-5)
4. Heat half pot of water, boil tortellini over medium for 5 minutes. Remove.
5. Heat the tomato sauce, cook the tortellini in the sauce for 1 minutes. Serve.

TIPS

- Avoid using too much filling for each tortellini; any air inside must be pressed out, and tortellini must be rolled tightly.

蛋奶素

青瓜意大利麵伴芥末蜜糖乳酪醬
Cucumber Noodles in Honey Yoghurt Sauce

| 材料 |

優質青瓜1個（大）
車打芝士1片（撕細塊）
松子仁1湯匙
芥末蜜糖乳酪醬（見p.104）

| 做法 |

1. 青瓜洗淨，切去頭尾兩端。
2. 青瓜以專用刨麵機，刨出青瓜意大利麵。
3. 青瓜意大利麵加入芥末蜜糖乳酪醬拌勻，上碟，鋪上車打芝士，灑入松子仁食用。

| INGREDIENTS |

1 large quality cucumber
1 slice cheddar cheese (tear into small pieces)
1 tbsp pine nuts
honey yoghurt sauce (see p.104)

| METHOD |

1. Rinse cucumber and cut off both ends.
2. Use a spiraliser to make cucumber noodles.
3. Mix the cucumber noodle with honey yoghurt sauce and put on the plate. Top with cheddar cheese and pine nuts. Serve.

煮素小貼士 

・建議選用本地有機青瓜或意大利青瓜，做出來的效果較好。

TIPS

• Local organic or Italian cucumber are recommended for best texture and taste.

蛋奶素

鮮露筍車厘茄芝士筆尖粉

Flaxseed Penne with Asparagus and Cherry Tomatoes

| 材料 |

鮮露筍2棵

車厘茄8粒

芝士1片（撕細塊）

蒜肉2粒（拍碎）

亞麻籽筆尖粉100克

海鹽1茶匙

| 調味料 |

黑椒碎半茶匙

| 做法 |

1. 鮮露筍削去硬外皮，洗淨，切斜段；車厘茄去蒂，洗淨。
2. 煮滾半鍋水，放入海鹽、亞麻籽筆尖粉煮約8分鐘，盛起（烚水留用）。
3. 平底鑊下橄欖油2湯匙，下蒜肉、鮮露筍炒香，加入亞麻籽筆尖粉、烚意粉水3湯匙、調味料及車厘茄煮片刻，加入芝士拌勻，上碟食用。

| INGREDIENTS |

2 fresh asparagus

8 cherry tomatoes

1 slice cheese (tear into small pieces)

2 cloves garlic (crushed)

100 g flaxseed penne

1 tsp sea salt

| SEASONING |

1/2 tsp finely chopped black pepper

| METHOD |

1. Peel off the hard skin from asparagus, rinse and cut into sections at an angle; remove stalks from cherry tomatoes and rinse.
2. Heat half pot of water. Add sea salt and flaxseed penne and cook for about 8 minutes. Remove and keep the water.
3. Heat wok and add 2 tbsp of oil. Stir-fry garlic and asparagus until fragrant. Add flaseed penne and 3 tbsp of water from boiling penne, seasoning and cherry tomatoes and cook for a while. Mix in cheese and serve.

煮素小貼士

· 市面出售的普通筆尖粉、螺絲粉、蝴蝶粉或通心粉皆可。

TIPS

- Normal penne, fusilli, farfalle and macaroni can be used as well.

蛋奶素

素千層麵
Vegetarian Lasagna

| 材料 |

番茄醬1碗（見p.22）

罐裝紅腰豆半碗

車打芝士絲100克

千層麵皮6至8塊

| 做法 |

1. 紅腰豆壓成茸，加入番茄醬拌勻。
2. 焗盤內鋪入一層番茄豆醬，灑下適量芝士絲，鋪上一塊千層麵皮，再加入一層番茄豆醬、芝士絲，直至蓋滿焗盤，最上層是番茄豆醬與芝士絲。
3. 焗爐預熱至180℃，放入素千層麵上下火焗30分鐘，待面層番茄豆醬呈金黃、芝士絲溶化即成。

| INGREDIENTS |

1 bowl tomato sauce (see p.22)

1/2 bowl canned kidney beans

100 g shredded cheddar cheese

6 to 8 pieces lasagna noodles

| METHOD |

1. Crush and mash kidney beans, mix well with tomato sauce.
2. In a baking tray, place a layer of tomato bean sauce and cheese, put a piece of lasagna noodle, repeat the layering until it is full, make sure the top layer is tomato bean sauce and cheese.
3. Preheat oven to 180°C. Bake the lasagna for 30 minutes, until the top layer browned and cheese dissolved. Serve.

煮素小貼士

・現時市售的千層麵皮毋須烚軟（購買時查閱包裝盒），只要依次鋪上餡料、芝士絲、麵皮，至蓋滿整個焗盤即可，簡單方便。

TIPS

- Some readymade lasagna noodles can be used right away without scalding (check the instructions on package). Making this lasagna is as simple as layering the filling, cheese and the noodles.

全素

素肉扒漢堡包
Vegetarian Hamburger

| 材料 |

素肉扒2塊（見p.26）
番茄1個
紅洋葱絲少許
沙律生菜適量
漢堡包2個

| 醬汁 |

欖油羅勒醬適量（見p.16）

| 做法 |

1. 番茄去蒂，洗淨，切片。
2. 漢堡包放入焗爐，用150℃焗10分鐘，保溫。
3. 素肉扒放入平底鑊，用小火慢慢烘至兩面金黃。
4. 生菜、紅洋葱、番茄、素肉扒排在漢堡包底層，淋上適量欖油羅勒醬，蓋上漢堡包面層即可。

| INGREDIENTS |

2 pieces vegetarian burger (see p.26)
1 tomato
shredded red onion
salad lettuces
2 hamburger buns

| SAUCE |

pesto sauce (see p.16)

| METHOD |

1. Remove the stalk from tomato, rinse and slice.
2. Roast hamburger buns in an oven at 150°C for 10 minutes, keep warm.
3. Roast vegetarian burgers in a pan over low heat, until both side browned.
4. Place lettuce, red onion, tomato, vegetarian burger on top of the base buns. Top with the sauce. Finish with the top bun. Serve.

煮素小貼士

- 加入冬菇的素肉扒，有嚼勁之餘，也有陣陣菇香味。

TIPS

- Vegetarian burgers are made from mushrooms, they have a chewy texture and great mushroom aroma.

全素

南瓜洋葱蘋果煮椰菜花
Braised Cauliflower and Pumpkin with Onion and Apple

| 材料 |

椰菜花12兩

蘋果1個

南瓜6兩

洋葱半個

蒜肉3粒（拍碎）

| 調味料 |

薑黃粉1茶匙

肉桂粉半茶匙

海鹽半茶匙

| 做法 |

1. 椰菜花切細朵，用水浸半小時，洗淨。
2. 蘋果洗淨外皮，去蒂、去芯，切塊。
3. 南瓜去籽，洗淨，切細塊。
4. 洋葱去外衣，洗淨，切碎。
5. 燒熱鑊下油2湯匙，下蒜肉、洋葱炒香，加入南瓜、蘋果及熱水3碗煮10分鐘，下椰菜花煮10分鐘，灑入調味料煮滾即成。

| INGREDIENTS |

450 g cauliflower
1 apple
225 g pumpkin
1/2 onion
3 cloves garlic (crushed)

| SEASONING |

1 tsp turmeric
1/2 tsp ground cinnamon
1/2 tsp sea salt

| METHOD |

1. Cut cauliflower into small pieces, soak for 30 minutes, rinse.
2. Rinse apple, remove stalk and core and cut into pieces.
3. Remove seeds from pumpkin, rinse and cut into small pieces.
4. Peel onion, rinse and chop.
5. Heat wok and add 2 tbsp of oil. Stir-fry garlic and onion until fragrant. Add pumpkin, apple and 3 bowls of hot water and cook for 10 minutes. Add cauliflower and cook for 10 minutes. Add seasoning and bring to boil. Serve.

煮素小貼士

- 可轉用西蘭花、塔花或紫色、橙色、綠色的椰菜花皆可。

TIPS

- Broccoli, Roman broccoli and purple, orange or green cauliflower can be used for this dish.

芝士烤鮮蔬
Roasted Vegetables with Cheese

蛋奶素

| 材料 |

大蘑菇2個
甜椒2個
翠玉瓜1條
茄子1條
車打芝士2片（切條）
海鹽1茶匙

| 做法 |

1. 大蘑菇用廚房紙抹淨；甜椒去蒂、去籽，洗淨，切大塊；翠玉瓜洗淨，切厚片；茄子去蒂，洗淨，切厚片。
2. 全部蔬菜材料加入海鹽拌勻，待10分鐘後排上網架，放入已預熱焗爐用180℃的上下火焗30分鐘，灑上芝士焗約10分鐘，待芝士開始溶化，取出放暖供食。

| INGREDIENTS |

2 portobello mushrooms
2 bell peppers
1 zucchini
1 eggplant
2 slices cheddar cheese (shredded)
1 tsp sea salt

| METHOD |

1. Wipe portobello mushrooms with kitchen paper; remove stalks and seeds from bell peppers, rinse and cut into large pieces; rinse zucchini, cut into thick slices; remove stalk from eggplant, rinse and cut into thick slices.
2. Mix all vegetables with sea salt and let it sit for 10 minutes. Arrange the vegetables on a baking tray. Bake in a preheated oven at 180°C for 30 minutes. Add shredded cheese and bake for 10 minutes, until it starts dissolving. Remove and let it warm. Serve.

煮素小貼士

- 蔬菜先與鹽拌勻，令鮮蔬帶鹹味，也令蔬菜釋出水分。

TIPS

- Salt can add salty flavour to the vegetables and release moisture from them.

惹味素菜
Vegetarian Dishes with Tangy Taste

全素

麻辣茄子豆腐蒟蒻結
Spicy Eggplant, Tofu and Konnyaku

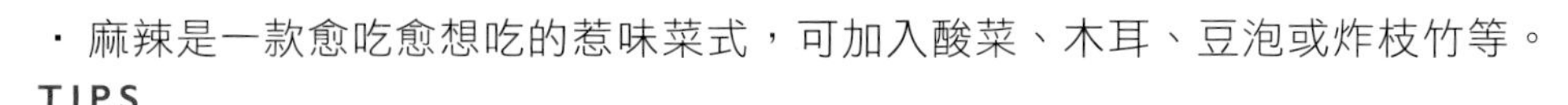

煮素小貼士

・麻辣是一款愈吃愈想吃的惹味菜式，可加入酸菜、木耳、豆泡或炸枝竹等。

TIPS

- Numbing and spicy dishes in Sichuan style is an addictive dish, you can also add pickled mustard green, black fungus, deep-fried tofu or deep-fried tofu stick.

| 材料 |

茄子1條
實板豆腐1塊
蒟蒻結1包
豆瓣醬2湯匙
紅辣椒2隻
乾辣椒2隻
蒜肉3粒（拍碎）
花椒油2茶匙

| 調味料 |

黃砂糖1茶匙
生抽2茶匙

| 做法 |

1. 茄子去蒂，洗淨，切段。
2. 板豆腐洗淨，切塊；蒟蒻結洗淨，瀝乾水分。
3. 燒熱鑊下油2湯匙，放入茄子、豆腐略煎，保溫。
4. 燒熱鍋下油2湯匙，下蒜肉、辣椒、豆瓣醬拌香，加入茄子、豆腐、乾辣椒及熱水1碗煮滾，再下蒟蒻結及調味料煮片刻，最後灑入花椒油煮滾即成。

| INGREDIENTS |

1 eggplant
1 cube firm tofu
1 pack konnyaku bundles
2 tbsp spicy bean paste
2 red chilli
2 dried red chilli
3 cloves garlic (crushed)
2 tsp Sichuan pepper oil

| SEASONING |

1 tsp brown sugar
2 tsp light soy sauce

| METHOD |

1. Remove stalk from eggplant, rinse and cut into sections.
2. Rinse tofu, cut into piece; rinse konnyaku bundles and drain.
3. Heat wok and add 2 tbsp of oil, stir-fry eggplant and tofu briefly. Keep warm.
4. Heat pot and add 2 tbsp of oil, stir-fry garlic, red chilli and spicy bean paste until fragrant. Add eggplant, tofu, dried red chilli and 1 bowl of hot water and bring to boil. Add konnyaku bundles and seasoning and cook. Sprinkle with Sichuan pepper oil and bring to boil. Serve.

全素

酸甜醬菠蘿伴豆卷
Tofu Rolls with Pineapple and Sweet Sour Sauce

| 材料 |

豆卷1盒
新鮮菠蘿1/4個
雪耳1/2球
三色甜椒各1/4個
乾葱2粒（去衣、切件）

| 酸甜汁料 |（調勻）

茄汁3湯匙
青檸汁2湯匙
糖2茶匙
鹽半茶匙
粟粉1茶匙
水半碗

| 做法 |

1. 雪耳用水浸1小時，去硬蒂，撕細塊，飛水，瀝乾水分。
2. 三色甜椒去籽，洗淨，切粒；菠蘿切小塊。
3. 燒熱鑊下油2湯匙，下乾葱拌香，傾下酸甜汁料煮滾，放入雪耳煮片刻，加入甜椒、菠蘿煮一會，上碟，伴豆卷蘸吃。

| INGREDIENTS |

1 box deep-fried tofu roll
1/4 fresh pineapple
1/2 white fungus
1/4 green bell pepper
1/4 red bell pepper
1/4 yellow bell pepper
2 shallots (peeled and cut)

| SWEET AND SOUR SAUCE | (MIXED WELL)

3 tbsp ketchup
2 tbsp lime juice
2 tsp sugar
1/2 tsp salt
1 tsp cornflour
1/2 bowl water

| METHOD |

1. Soak white fungus for 1 hour, cut off hard stems and tear into small pieces. Scald and drain.
2. Remove seeds from the bell peppers, rinse and dice; cut pineapple into small pieces.
3. Heat wok and add 2 tbsp of oil. Sir-fry shallots until fragrant, add sweet and sour sauce and bring to boil. Add white fungus and cook. Add bell peppers and pineapple and cook. Plate and serve with deep-fried tofu roll.

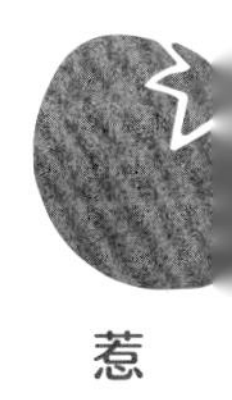

煮素小貼士

・酸甜醬可配搭炸素雲吞、油條進食，口感也非常好。

TIPS

- Sweet and sour sauce can be served with deep-fried vegetarian dumplings or Chinese sticky sticks, the texture is as good.

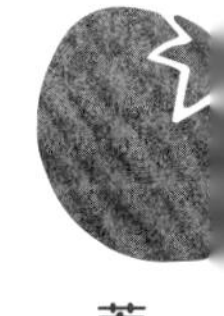

全素

辣椒蒜茸炒龍鬚菜
Stir-fried Chayote Shoots with Chilli and Garlic

| 材料 |

新鮮龍鬚菜12兩（合掌瓜苗）
乾辣椒2隻
蒜茸2茶匙

| 調味料 |

黃砂糖半茶匙
海鹽半茶匙

| 做法 |

1. 龍鬚菜撕去老筋，摘掉菜鬚，洗淨，切段。
2. 燒熱鑊下油2湯匙，下乾辣椒、蒜茸炒香，加入龍鬚菜炒勻，下熱水3湯匙及調味料再炒片刻即可。

| INGREDIENTS |

450 g fresh chayote shoots
2 dried red chillis
2 tsp grated garlic

| SEASONING |

1/2 tsp brown sugar
1/2 tsp sea salt

| METHOD |

1. Tear off hard veins from chayote shoots, rinse and cut into sections.
2. Heat wok and add 2 tbsp of oil. Stir-fry dried red chilli and garlic until fragrant. Add chayote shoots and stir-fry. Add 3 tbsp of hot water and seasoning and stir fry. Serve.

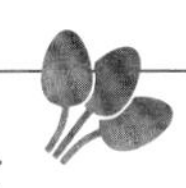

煮素小貼士

- 炒後的龍鬚菜脆嫩好吃，如街市有苦瓜苗、南瓜苗也可一試，同樣好吃。
- 乾辣椒令餸菜香濃惹味，但不辣。

TIPS

- Stir-fried chayote shoots are crunchy and delicious, same goes for bitter melon and pumpkin shoots.
- Dried red chilli enhance the flavour of the dish, without making it spicy.

全素

綠咖喱銀杏豆泡煮綠椰菜花

Green Curry with Green Cauliflower, Deep-fried Tofu Puffs and Ginkgo

煮素小貼士

・如買不到綠色椰菜花，普通白色椰菜花也可。

TIPS

- You can use normal white cauliflower if green ones are not available.

材料

綠色椰菜花8兩
四季豆4兩
銀杏2兩
豆泡3兩
香茅2條
金不換1棵（洗淨）
乾葱3粒（去衣、切片）
綠咖喱醬4湯匙
椰漿3湯匙

調味料

椰糖1湯匙
鹽半茶匙

做法

1. 綠椰菜花切細朵，用水浸半小時，洗淨。
2. 四季豆撕去老筋，洗淨，摘短度。
3. 銀杏飛水，去外衣，洗淨。
4. 豆泡飛水，過冷河，擠乾水分。
5. 燒熱鑊下油2湯匙，下乾葱炒香，加入綠咖喱醬、銀杏、四季豆、香茅炒勻，下熱水3碗煮滾，加入綠椰菜花、豆泡、調味料煮片刻，最後下椰漿慢火煮至微滾，灑入金不換即可。

INGREDIENTS

300 g green cauliflower
150 g snap beans
75 g ginkgo
113 g deep-fried tofu puffs
2 stalks lemongrass
1 stalk Thai basil (rinsed)
3 shallots (peeled, sliced)
4 tbsp green curry paste
3 tbsp coconut milk

SEASONING

1 tbsp palm sugar
1/2 tsp salt

METHOD

1. Cut green cauliflower into small pieces, soak for 30 minutes, rinse.
2. Tear off hard veins from snap beans, rinse and tear into short sections.
3. Rinse ginkgo, peel and rinse.
4. Scald deep-fried tofu puffs, rinse with cold water and squeeze until dry.
5. Heat wok and add 2 tbsp of oil, stir-fry shallots until fragrant. Add green curry paste, ginkgo, snap beans and lemongrass and stir fry. Add 3 bowls of hot water and bring to boil. Add green cauliflower, deep-fried tofu puffs and seasoning and cook. Add coconut milk and simmer over low heat. Top with Thai basil and serve.

全素

酸辣老鼠耳
Sour and Spicy Black Fungus

| 材料 |

小朵雲耳半兩

芫茜2棵

紅辣椒2隻

蒜茸2茶匙

| 調味料 |

陳醋3湯匙

生抽半湯匙

黃砂糖2茶匙

麻油2茶匙

| 做法 |

1. 小朵雲耳用水浸2小時，飛水，過冷河，瀝乾水分。
2. 芫茜切去鬚根，洗淨，切碎；紅辣椒洗淨，去蒂，切粒。
3. 雲耳加入蒜茸、辣椒、調味料拌勻，冷藏2小時，加入芫茜拌勻即可。

| INGREDIENTS |

19 g mini black fungus
2 stalks coriander
2 red chillies
2 tsp grated garlic

| SEASONING |

3 tbsp aged vinegar
1/2 tbsp light soy sauce
2 tsp brown sugar
2 tsp sesame oil

| METHOD |

1. Soak mini black fungus for 2 hours, scald, rinse with cold water and drain.
2. Cut off the roots from coriander, rinse and chop; rinse red chilli, remove stalks and dice.
3. Mix black fungus, garlic, red chillies and seasoning together. Refrigerate for 2 hours. Mix in coriander and serve.

煮素小貼士

・酸辣小雲耳是夏日的開胃菜式，亦可改用木耳切絲，拌入調味料皆美味。

TIPS

- This dish is a perfect summer appetiser. You can also use shredded wood ear fungus as well.

全素

韓式泡菜炒粉絲
Stir-fried Sweet Potato Noodles in Korean Style

| 材料 | （4人份量）

韓式粉絲80克
泡菜1小碟
青、紅甜椒各適量
鮮菇適量
洋葱1/4個

| 調味料 |

生抽半湯匙
韓式甜辣醬半湯匙
麻油2茶匙
水半碗

| 做法 |

1. 韓式粉絲用水浸1小時，放入熱水浸片刻，撈起，瀝乾水分。
2. 甜椒去籽，洗淨，切絲；鮮菇洗淨，瀝乾水分；洋葱去外衣，洗淨，切絲。
3. 燒熱鑊下油2湯匙，下洋葱炒香，加入鮮菇、甜椒及泡菜炒勻，下調味料煮滾，加入韓式粉絲拌勻略煮，即可上碟。

| INGREDIENTS | (4 SERVINGS)

80 g Korean sweet potato starch noodles
1 small plate kimchi
green bell pepper
red bell pepper
fresh mushrooms
1/4 onion

| SEASONING |

1/2 tbsp light soy sauce
1/2 tbsp Korean red chilli paste
2 tsp sesame oil
1/2 bowl water

| METHOD |

1. Soak Korean noodles for 1 hour, soak in hot water, remove and drain.
2. Remove seeds from bell peppers, rinse and shred; rinse mushrooms and drain; peel onion, rinse and shred.
3. Heat wok and add 2 tbsp of oil. Stir-fry onion until fragrant, add mushrooms, bell peppers and kimchi and stir well. Add seasoning and bring to boil. Mix in Korean noodle and cook briefly. Serve.

煮素小貼士

・韓式粉絲建議最後才加入炒煮，否則容易糊成一糰。

TIPS

- Korean noodles is recommended to add last, or it will dissolve and stick together.

全素

蒸茄子伴陳醋汁
Steamed Eggplant with Vinegar Sauce

| 材料 |

茄子12兩
紅辣椒3隻
芫茜2棵
炒香白芝麻2茶匙

| 汁料 |

陳醋1湯匙
老抽半湯匙
黃砂糖1茶匙

| 做法 |

1. 茄子去蒂，洗淨，切段，隔水大火蒸10分鐘，傾去汁液，備用。
2. 芫茜去鬚根，洗淨，切碎。
3. 紅辣椒去蒂，洗淨，切粒，放入熱油拌香，加入汁料內拌勻，淋在茄子上，最後灑下芫茜及白芝麻。

| INGREDIENTS |

450 g eggplant
3 red chillies
2 stalks coriander
2 tsp toasted sesame

| SAUCE |

1 tbsp aged vinegar
1/2 tbsp dark soy sauce
1 tsp brown sugar

| METHOD |

1. Remove stalk from eggplant, rinse and cut into sections. Steam over high heat for 10 minutes, discard the juice, set aside.
2. Cut off roots from coriander, rinse and chop.
3. Remove stalks from red chillies, rinse and dice. Stir-fry red chilli with oil until fragrant. Add the sauce and mix well, pour the mixture on top of eggplant. Top with coriander and sesame. Serve.

煮素小貼士

- 蒸吃的茄子宜選購幼身長形的，蒸好的茄子較軟滑。

TIPS

- Choose thin and long eggplants for steaming; they are softer and more smooth.

紅咖喱煮秋葵薯仔
Red Curry with Potato and Okra

| 材料 |

馬鈴薯1斤
草菇10粒
秋葵8條（去蒂）
乾葱3粒（去衣、切片）
香茅2條
檸檬葉4塊
紅咖喱醬4湯匙
椰絲3湯匙

| 調味料 |

黃砂糖1茶匙
鹽半茶匙

| 做法 |

1. 馬鈴薯去皮，洗淨，切滾刀塊，加水浸5分鐘，瀝乾水分。
2. 草菇洗淨，飛水，過冷河，瀝乾水分。
3. 燒熱鑊下油2湯匙，下乾葱炒香，加入馬鈴薯、紅咖喱醬、香茅、檸檬葉炒勻，下熱水4碗煮滾，轉小火煮20分鐘，加入草菇、秋葵、調味料及椰絲拌勻，煮滾片刻即可。

| INGREDIENTS |

600 g potatoes
10 straw mushrooms
8 okra (remove stalks)
3 shallots (peeled, sliced)
2 stalks lemongrass
4 lime leaves
4 tbsp red curry paste
3 tbsp coconut flakes

| SEASONING |

1 tsp brown sugar
1/2 tsp salt

| METHOD |

1. Peel potatoes, rinse and cut an angle into pieces. Soak in water for 5 minutes, drain.
2. Rinse straw mushrooms, scald for a while. Rinse with cold water and drain.
3. Heat wok and 2 tbsp of oil. Stir-fry shallots until fragrant. Add potatoes, red curry paste, lemongrass, lime leaves and stir fry. Add 4 bowls of hot water and bring to boil. Turn to low heat and simmer 20 minutes. Add straw mushrooms, okra, seasoning and coconut flakes and mix well. Bring to boil. Serve.

煮素小貼士

- 可用紅咖喱煮黃心番薯，味道更甜美。
- 今次改以椰絲代替椰漿，煮出來更濃、更香。

TIPS

- Yellow sweet potatoes can be cooked with red curry as well.
- Using coconut flakes give a strong taste and aroma, instead of coconut milk.

全素

酸菜木耳燜枝竹
Braised Tofu Stick with Wood Ear Fungus and Pickled Mustard Green

| 材料 |

炸枝竹3條
鹹酸菜4兩
木耳1大塊
薑3片
芝麻醬1湯匙

| 調味料 |

麵豉醬半湯匙
黃砂糖1茶匙

| 做法 |

1. 木耳用水浸1小時，飛水，過冷河，切幼條。
2. 鹹酸菜洗淨，切幼條。
3. 炸枝竹折成短度，飛水，過冷河，擠乾水分。
4. 燒熱鑊下油2湯匙，下薑片炒香，加入木耳、鹹酸菜炒勻，拌入熱水2碗及調味料煮滾，下炸枝竹用小火煮15分鐘，拌入芝麻醬，上碟即可。

| INGREDIENTS |

3 deep-fried tofu sticks
150 g pickled mustard green
1 large wood ear fungus
3 slices ginger
1 tbsp sesame paste

| SEASONING |

1/2 tbsp soybean paste
1 tsp brown sugar

| METHOD |

1. Soak wood ear fungus in water for 1 hour, scald for a while. Rinse with cold water and shred finely.
2. Rinse pickled mustard green, shred finely.
3. Tear deep-fried tofu stick into short sections. Scald, rinse with cold water and squeeze until dry.
4. Heat wok and add 2 tbsp of oil. Stir-fry ginger until fragrant, add wood ear fungus and pickled mustard green and stir-fry. Add 2 bowls of hot water and seasoning and bring to boil. Add deep-fried tofu sticks and cook over low heat for 15 minutes. Mix in sesame paste. Serve.

煮素小貼士

- 如鹹酸菜太鹹，可用鹽1茶匙加適量清水，浸鹹酸菜15分鐘，讓鹽水帶走鹹味，再洗淨。

TIPS

- Soaking the pickled mustard green in the 1 tsp of salted water for 15 minutes, it will expel the saltiness. Then rinse the pickled mustard green again before cooking.

蛋奶素

葡汁甘筍燴炸豆腐
Braised Tofu and Carrot in Portuguese Sauce

| 材料 |

炸豆腐4塊
甘筍1條（去皮、切片）
西蘭花適量
乾葱3粒（去衣、切片）
椰漿1杯
忌廉2湯匙
咖喱粉1茶匙
黃薑粉2湯匙

| 調味料 |

黃砂糖1茶匙
鹽3/4茶匙

| 做法 |

1. 炸豆腐用熱水洗去表面油分，抹乾水分，切細塊。
2. 西蘭花切細朵，用水浸半小時，洗淨，飛水。
3. 燒熱鑊下油2湯匙，下乾葱炒香，下甘筍、咖喱粉、黃薑粉炒勻，加入熱水及椰漿各1杯煮滾，下炸豆腐、調味料煮滾片刻，加入忌廉用慢火邊煮邊攪拌，最後下西蘭花略煮即可。

| INGREDIENTS |

4 pieces deep-fried tofu
1 carrot (peeled, sliced)
broccoli
3 shallots (peeled, sliced)
1 cup coconut milk
2 tbsp whipping cream
1 tsp curry powder
2 tbsp turmeric powder

| SEASONING |

1 tsp brown sugar
3/4 tsp salt

| METHOD |

1. Rinse deep-fried tofu with hot water. Wipe dry and cut into small pieces.
2. Cut broccoli into small pieces, soak in water for 30 minutes, rinse and scald.
3. Heat wok and add 2 tbsp of oil. Stir-fry shallots until fragrant, add carrot, curry powder and turmeric powder and stir fry. Add 1 cup of water and 1 cup of coconut milk and bring to boil. Add deep-fried tofu and seasoning and cook for a while. Turn to low heat and stir in whipping cream. Add broccoli and cook. Serve.

煮素小貼士

- 加入忌廉令汁液濃稠，如不喜歡可刪掉；黃薑粉可轉用薑黃粉。

TIPS

- Whipping cream can thicken the sauce, omit it as desired.

素味甜品、飲料
Vegetarian Desserts and Drinks

全素

蓮子百合杞子桂花桃膠糖水
Sweet Soup with Peach Gum, Lotus Seeds and Lily Bulbs

| 材料 |

桃膠2湯匙
蓮子1兩
百合1兩
杞子1湯匙
桂花2茶匙
冰糖適量

| 做法 |

1. 桃膠用水浸4至5小時，挑去污物，洗淨，飛水，瀝乾水分。
2. 蓮子去芯、洗淨，用水浸1小時，瀝乾水分。
3. 百合洗淨；杞子用熱水沖洗乾淨，隔去水分。
4. 煲滾清水6碗，放入蓮子煲半小時，加入百合、桃膠再煲15分鐘，下冰糖煲至糖溶化，最後加入杞子、桂花煲滾即成。

| INGREDIENTS |

2 tbsp peach gum
38 g lotus seeds
38 g lily bulbs
1 tbsp Qi Zi
2 tsp dried osmanthus
rock sugar

| METHOD |

1. Soak peach gum for 4 to 5 hours, pick out any dirts. Rinse, scald and drain.
2. Remove the core from lotus seeds. Rinse lotus seeds, soak for 1 hour and drain.
3. Rinse lily bulbs; rinse Qi Zi with hot water, drain.
4. Bring 6 bowls of water to boil. Boil lotus seeds for 30 minutes. Add lily bulbs and peach gum and boil for 15 minutes. Add rock sugar and boil until dissolved. Add Qi Zi and osmanthus, bring to boil. Serve.

煮素小貼士

- 此糖水具美顏、滋潤皮膚的功效。
- 桃膠摻有樹木雜質，細心地挑去污物，乾淨衛生。

TIPS

- This sweet soup can freshen and nourish your skin.
- Peach gum is mixed with dirts from the tree, they must be picked out carefully.

全素

楓糖漿拌焗番薯
Baked Sweet Potato with Maple Syrup

| 材料 |

黃心番薯2個

楓葉糖漿適量

| 做法 |

1. 黃心番薯洗淨，用錫紙包好。
2. 焗爐用150℃預熱10分鐘，放入番薯焗45分鐘。
3. 番薯取出，切滾刀塊，淋上適量楓葉糖漿，再焗5分鐘，待呈焦糖香味即可。

| INGREDIENTS |

2 yellow sweet potatoes

maple syrup

| METHOD |

1. Rinse yellow sweet potatoes and wrap with aluminium foil.
2. Preheat oven at 150°C for 10 minutes. Bake sweet potatoes for 45 minutes.
3. Cut sweet potatoes at an angle into pieces. Top with maple syrup and bake for 5 minutes until fragrant. Serve.

煮素小貼士

- 番薯纖維含量高，有助排便，潔淨腸道，改善皮膚。
- 錫紙反光面向內包好，可加速食物熟透。

TIPS

- Sweet potatoes are rich in fibre; they enhance digestive system, clean the bowel and improve skin condition.
- Use the shiny side of the aluminium on the inside, touching the sweet potatoes; they will be faster to cook.

全素

紅棗栗子冰糖燉黃耳
Stewed Yellow Fungus with Red Dates and Chestnuts

| 材料 |

黃耳1/3兩

新鮮栗子4兩

紅棗3粒

冰糖適量

| 做法 |

1. 黃耳用水浸5至6小時，洗淨，切細朵。
2. 栗子飛水，去衣，洗淨；紅棗去核，剪成粗條。
3. 全部材料放入燉盅，注入滾水3碗，加蓋，用大火煲滾水10分鐘，轉小火燉1.5小時即成。

煮素小貼士

- 黃耳含豐富的膠質，常吃可補充骨膠原，兼具美顏功效。
- 黃耳必須長時間浸泡至軟身，否則質感較硬。

TIPS

- Yellow fungus are rich in collagen, it can supply collagen to your body and nourish your skin.
- Yellow fungus should be soaked for a long time until soft, or it will have a hard texture.

| INGREDIENTS |

13 g yellow fungus
150 g fresh chestnuts
3 red dates
rock sugar

| METHOD |

1. Soak yellow fungus for 5-6 hours, rinse and cut into small pieces
2. Scald chestnuts, peel and rinse; core red dates and cut into strips.
3. Put all ingredients in a stewing pot, add 3 bowls of boiling water. Cover the stewing pot. Boil the stewing pot over high heat for 10 minutes, turn to low heat and simmer for 1.5 hours. Serve.

全素

南瓜凍糕
Chilled Pumpkin Cake

| 材料 |

日本南瓜8兩
椰糖2湯匙
木薯粉75克（約6湯匙）
椰絲4湯匙
水1.5碗

| 做法 |

1. 南瓜去籽，去皮，切塊，隔水蒸15分鐘至全熟，趁熱搓成南瓜茸。
2. 椰糖及水1.5碗煮至椰糖溶化，備用。
3. 南瓜茸及木薯粉拌勻，下椰糖水慢慢拌成幼滑南瓜漿。
4. 南瓜漿傾進容器，隔水蒸半小時，待冷，冷藏2小時，切塊，均勻地滾上椰絲即成。

| INGREDIENTS |

300 g pumpkin
2 tbsp palm sugar
75 g cassava flour (about 6 tbsp)
4 tbsp coconut flakes
1.5 bowls water

| METHOD |

1. Remove seeds and skin from pumpkin and cut into pieces. Steam for 15 minutes until cooked and mash while still hot.
2. Heat 1.5 bowls of water and palm sugar together until the sugar dissolves.
3. Mix mashed pumpkin and cassava flour together, add palm sugar mixture and stir until smooth.
4. Pour the pumpkin mixture into containers and steam for 30 minutes. Let it cool and refrigerate for 2 hours. Cut the pumpkin cake into pieces and coat with coconut flakes. Serve.

煮素小貼士

・建議選用日本南瓜，南瓜糕才味香、質感軟綿。

TIPS

• Japanese pumpkins are recommended as they have stronger aroma and softer texture.

蛋奶素

燕麥腰果奶
Oatmeal and Cashew Milk

| 材料 |

烘熱腰果20粒

熟燕麥3湯匙

鮮奶1杯

凍開水1杯

| 做法 |

全部材料放入攪拌機，磨成幼滑的燕麥腰果奶即可。

| INGREDIENTS |

20 roasted cashew nuts
3 tbsp cooked oatmeal
1 cup milk
1 cup cold boiled water

| METHOD |

Blend all ingredients in a blender into a smooth milk. Serve.

煮素小貼士

- 全素食者可減去鮮奶份量，味道同樣美味。
- 燕麥腰果奶營養豐富，幫助腸道有利排便。
- 飲用時可拌入胚芽，吸收維他命E，促進血液流通，抗氧化。

TIPS

- Vegan can omit the milk and enjoy the same delicious drink.
- Oatmeal and cashew milk is very nutritious, it promotes bowel movement.
- Wheat germ can be added to the drink, its vitamin E can improve circulation and it is a great antioxidant.

杏仁蛋白茶
Egg White and Almond Drink

蛋奶素

| 材料 | （4人份量）

南杏3兩
北杏1湯匙
粘米3湯匙
蛋白2隻
冰糖1湯匙
水3碗

| 做法 |

1. 南杏、北杏、粘米同洗淨，加入水3碗浸2小時。
2. 將南北杏、粘米及水放入攪拌機，磨成幼滑杏仁漿，過濾。
3. 杏仁漿用中小火煮滾，加入冰糖用小火煮約10分鐘，熄火，最後拌入蛋白攪勻即可。

煮素小貼士

- 此甜品潤肺、美容、滋潤腸胃。
- 熄火後才拌入蛋白，質感嫩滑。

TIPS

- This dessert nourishes the Lung, skin and the digestive system.
- To make the smoothest drink, egg whites should be added after turning off heat.

| INGREDIENTS | (4 SERVINGS)

113 g sweet almonds
1 tbsp bitter almonds
3 tbsp white rice
2 egg whites
1 tbsp rock sugar
3 bowls water

| METHOD |

1. Rinse sweet and bitter almonds and rice together. Soak under 3 bowls of water for 2 hours.
2. Blend sweet and bitter almonds, rice and water in a blender together until smooth. Strain.
3. Bring the almond milk to boil over medium low heat. Add rock sugar and cook over low heat for 10 minutes. Turn off heat, add egg whites and stir well.

全素

苦瓜三青汁

Bitter Melon, Green Apple and Cucumber Juice

| 材料 |

苦瓜1個（小）

青蘋果2個

青瓜1條

凍開水1杯

蜜糖2茶匙

| 做法 |

1. 苦瓜切開去籽，洗淨，切塊。
2. 青蘋果洗淨，去蒂、去籽，切塊。
3. 青瓜洗淨，切去頭尾兩端，切塊。
4. 將苦瓜、青蘋果、青瓜放入榨汁機壓汁去渣，加入凍開水、蜜糖調勻，即可飲用。

| INGREDIENTS |

1 small bitter melon

2 Granny Smith apples

1 cucumber

1 cup cold boiled water

2 tsp honey

| METHOD |

1. Cut open bitter melon and remove seeds. Rinse and cut into pieces.
2. Rinse Granny Smith apples, remove stalks and seeds. Cut into pieces.
3. Rinse cucumber, cut off both ends and cut into pieces.
4. Juice bitter melon, Granny Smith apples and cucumber with a juicer. Add cold boiled water and honey and mix well. Serve.

煮素小貼士

- 這是三高人士的最佳飲品，唯血糖高者必須減去蜜糖為宜。
- 若怕太寒涼，建議加入薑2片同榨汁，或拌入薑粉半茶匙。

TIPS

- This is the best drink for people who have high level of blood pressure, sugar and lipid. However people with high blood sugar should omit the honey.
- If you cannot take this Cold drink, you can add 2 slices of ginger in the juicer or mix in 1/2 tsp of ginger powder in the drink.

全素

火龍果奇異果汁
Dragon Fruit and Kiwi Juice

| 材料 |（2杯份量）

紅肉火龍果1杯

奇異果1個

凍開水3/4杯

| 做法 |

1. 紅肉火龍果切細塊。
2. 奇異果去皮，切細塊。
3. 全部材料放入攪拌機，磨成幼滑火龍果奇異果汁，即可飲用。

| INGREDIENTS | (2 CUPS)

1 cup red dragon fruit
1 kiwifruit
3/4 cup cold boiled water

| METHOD |

1. Cut red dragon fruit flesh into small pieces.
2. Cut off skin from kiwifruit and cut into small pieces.
3. Blend all ingredients in a blender until smooth. Serve.

煮素小貼士

- 可補充維他命C，有助通腸排便；但需留意紅肉火龍果的果糖高，一天不可飲用太多，尤其血糖高者慎用。

TIPS

- This drink supplies with vitamin C and promotes bowel movements. Note that red dragon fruit is rich in sugar, do not drink too much in a day, especially people with high blood sugar.

全素

玫瑰花杞子圓肉茶
Rose, Qi Zi and Longan Tea

| 材料 |

有機大朵紅玫瑰2朵
杞子20粒
圓肉12粒

| 做法 |

1. 杞子、圓肉放入熱水洗淨，隔去水分。
2. 全部材料放入茶壺或保溫壺，沖入滾水2杯，加蓋焗15分鐘即可。

| INGREDIENTS |

2 large organic dried rose buds
20 Qi Zi
12 dried longan

| METHOD |

1. Rinse Qi Zi and dried longan in hot water, drain.
2. Put all ingredients in a tea pot or thermos bottle. Add 2 cups of hot water and sit for 15 minutes. Serve.

煮素小貼士

- 此茶疏肝明目、解鬱。
- 挑選大顆及色澤較紅的杞子。

TIPS

- This tea nourishes the Liver, improve eyesight and relieve stagnation.
- Choose large and red Qi Zi.

全素

紅棗南棗黨參茶
Red Date, Black Date and Dang Shen Tea

| 材料 |

黨參2條

南棗6粒

紅棗3粒

| 做法 |

1. 黨參洗淨，切斜段。
2. 紅棗去核，剪粗條；南棗洗淨。
3. 全部材料放入煲，注入清水3.5杯煲滾，轉小火煲45分鐘，傾入保溫瓶保溫，可全日飲用。

| INGREDIENTS |

2 stalks Dang Shen
6 black dates
3 red dates

| METHOD |

1. Rinse Dang Shen and cut at an angle into sections.
2. Core red dates and cut into strips; rinse black date.
3. Put all ingredients in a pot. Add 3.5 cups of water and bring to boil. Turn to low heat and boil for 45 minutes. Transfer into a thermos bottle to keep it hot for a whole day.

煮素小貼士

- 女性月事完畢，飲用此茶可補氣血，滋補身體。
- 黨參切成斜段焗飲，更能釋出味道及功效。

TIPS

- This tea benefits the Qi, blood and your body, it is suitable for drinking after menstruation.
- The most flavour and effect can be achieved when Dang Shen is cut at an angle.

木敦果茶
Bael Fruit Tea

全素

| 材料 |

木敦果片4片

楓葉糖漿2湯匙

| 做法 |

1. 木敦果片用清水洗淨。
2. 木敦果片放入煲，加水6杯煲滾，轉中火煲15分鐘，熄火待10分鐘，盛起，加入適量楓葉糖漿拌飲。

| INGREDIENTS |

4 slices dried bael

2 tbsp maple syrup

| METHOD |

1. Rinse bael slices.
2. Put bael slices and 6 cup of water in a pot. Bring to boil, turn to medium heat and boil for 15 minutes. Turn off heat and remain covered for 10 minutes. Transfer and mix with maple syrup. Serve.

煮素小貼士

- 木敦果茶可調節腸胃，幫助消化，通腸助大便。
- 炎熱的夏天，可煲多些果茶冷藏，清熱解暑。
- 加蓋待10分鐘，有效令木敦果散出香味。

TIPS

- Bael fruit tea can regulate and improve digestive system.
- In hot summer, prepare more bael fruit tea and refrigerate.
- After turning off heat, remain it covered to let more flavour release from the dried bael.

增強體質，抵抗都市病
Strengthen your Body to Resist Diseases of Affluence

素食有益身體，大概是常識吧！

懂得選、懂得吃，透過大自然的食材，可以強骨骼、增記憶、降膽固醇、防高血壓，從蔬菜、果仁飲食改善三高及其他都市健康問題。

It is common sense that vegetarian is beneficial to our body and health!

When you know how to choose and use natural vegetarian ingredients, you can strengthen your bone, memory functions, lower cholesterol and blood pressure. A diet with vegetables and nuts alleviates high level of blood lipid, blood sugar, pressure and other diseases of affluence.

補鈣強骨骼

代表食材：海帶、黑豆、白芝麻、菠菜

海藻類含鈣量豐富，特別是海帶，煮湯或麵不妨加點伴吃；白芝麻的鈣質含量也高，建議上碟時灑上烘香芝麻；菠菜也含鐵質，具補血功效。發育中的小朋友及銀髮一族，多吃鈣質食物增強骨質及牙齒健康。

CALCIUM RICH AND BONE STRENGTHENING

Notable ingredients: kelp, black soybeans, white sesame, spinach

Seaweeds are rich in calcium, especially kelps, add some to soups and noodles; sesame also is rich in calcium, top any dishes with toasted sesame; spinach contains calcium and also iron, which can reinvigorate blood. Children and elderly should absorb more calcium to keep their bones and teeth healthy.

補腦增記憶

代表食材：薑黃、合桃、銀杏

薑黃、合桃、銀杏等具補腦及健腦的功能，有效幫助腦部減壓，增強記憶，其中以薑黃的功效顯著，可預防腦退化，日常可沖調薑黃水飲用，或灑入薑黃粉伴吃。

BRAIN AND MEMORY FUNCTION ENHANCING

Notable ingredients: turmeric, walnut, ginkgo

Turmeric, walnut and ginkgo all enhance our brain and memory functions and release pressure from our brain.Turmeric is the most effective of the three, it prevents dementia. Mix turmeric powder in water as a drink or add to any everyday dishes.

飲食注意事項

痛風人士：蝦、蟹、蠔等含嘌呤；菠菜、西蘭花及鮮露筍含草酸；內臟、骨髓、肥油厚膏含大量脂肪、膽固醇，少吃為宜。

糖尿病人士：實行低油、低鹽、低糖飲食，注意澱粉質攝入量，多喝水為宜。

DIET CAUTIONS

Gout patient: shrimps, crabs and oysters contain purine; spinach, broccoli, asparagus contain oxalic acid; animal organs, bone marrow and fat meats contain large amount of fat and cholesterol; avoid having these ingredients.

Diabetes patient: implement a low fat, low salt and low sugar diet, be aware of starch intake, drink more water.

防病護心、降膽固醇、強化血管
For Healthy Heart, Lowering Cholesterol, Strengthening Blood Vessel

蒜片炒小椰菜
Stir-fried Brussels Sprouts with Garlic

全素

| 材料 |

小椰菜8兩
甘筍半條
蒜肉3粒（去衣、切片）

| 調味料 |

黃砂糖1/4茶匙
海鹽1/3茶匙

| 做法 |

1. 小椰菜洗淨，切成兩半；甘筍去皮，洗淨，切條。
2. 小椰菜、甘筍放於蒸碟，隔水大火蒸5分鐘，隔去汁液，備用。
3. 取平底鍋下橄欖油2湯匙，下蒜片炒香，加入小椰菜、甘筍拌勻，下熱水2湯匙、調味料炒片刻，盛起即可。

| INGREDIENTS |

300 g brussels sprouts
1/2 carrot
3 cloves garlic (peeled, sliced)

| SEASONING |

1/4 tsp brown sugar
1/3 tsp sea salt

| METHOD |

1. Rinse Brussels sprouts and cut in half; peel carrot, rinse and cut into strips.
2. Place Brussels sprouts and carrots on a plate. Steam over high heat for 5 minutes, discard the juice. Set aside.
3. Heat pan and add 2 tbsp of olive oil. Stir-fry garlic until fragrant. Add Brussels sprouts and carrot and stir-fry. Add 2 tbsp of hot water and seasoning, stir-fry. Serve.

煮素小貼士

- 小椰菜產地的來源眾多，其中以泰國的價錢比較相宜。

TIPS

- Brussels sprouts come from various countries, those from Thailand are relatively cheap.

全素

欖油羅勒醬雙色小扁豆
Green and Orange Lentils in Pesto Sauce

| 材料 |

綠色小扁豆50克
橙色小扁豆50克
牛油果1個
醋醃紅菜頭適量
欖油羅勒醬適量（見p.16）

| 做法 |

1. 雙色小扁豆分別洗淨，放進蒸碟，隔水大火蒸10至15分鐘至全熟，放涼備用。
2. 牛油果切片；醋醃紅菜頭切粗粒。
3. 雙色小扁豆加入欖油羅勒醬拌勻，排上紅菜頭、牛油果食用。

| INGREDIENTS |

50 g green lentils
50 g orange lentils
1 avocado
pickled beetroot
pesto sauce (see p.16)

| METHOD |

1. Rinse green and orange lentils and arrange on a plate. Steam over high heat for 10-15 minutes until fully cooked. Let it cool.
2. Slice avocado; dice pickled beetroot.
3. Mix the lentils with pesto sauce. Top with beetroot and avocado Serve.

煮素小貼士

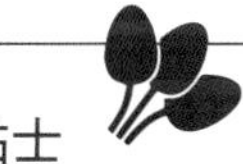

- 醋醃紅菜頭可自家製作：紅菜頭去皮，切厚塊，蒸熟後隔出汁液，待紅菜頭冷卻，加入蘋果醋，冷藏一星期可食。
- 小扁豆可早一晚蒸熟，放雪櫃備用。
- 紅菜頭隔出的汁液，可飲用；浸完紅菜頭的醋，可加水飲用，有益腸胃，幫助消化；但胃酸過多者慎用。

TIPS

- You can make your own pickled beetroot: peel beetroot and cut into thick slices. Steam until cooked, strain the juice and let it cool. Soak beetroot in apple vinegar and refrigerate for 1 week.
- Lentils can be steamed one night before, set aside in refrigerator.
- You can drink the strained juice from steaming beetroot; mix the vinegar from pickling beetroot with water, it is beneficial to digestive system. However, people with acid reflux should take cautions.

全素

鮮露筍拌素鴨絲
Stir-fried Tofu Skin and Asparagus

| 材料 |

鮮露筍4條（大）

素鴨1塊

乾葱2粒（去衣、切片）

| 調味料 |

素蠔油半湯匙

| 做法 |

1. 鮮露筍削去硬皮，洗淨，切斜段；素鴨切絲。
2. 燒熱鑊下橄欖油2湯匙，下乾葱拌香，加入鮮露筍炒勻，下素鴨、熱水2湯匙炒片刻，加入調味料炒勻，上碟供食。

| INGREDIENTS |

4 large asparagus
1 piece fried tofu skin (vegetarian duck)
2 shallots (peeled, sliced)

| SEASONING |

1/2 tbsp vegetarian oyster sauce

| METHOD |

1. Peel off the hard skin from asparagus, rinse and cut into sections at an angle; shred fried tofu skin.
2. Heat wok and add 2 tbsp of olive oil. Stir-fry shallots until fragrant, add asparagus and stir fry. Add fried tofu skin and 2 tbsp of hot water, stir-fry. Add seasoning and stir-fry. Serve.

煮素小貼士 

・如買不到粗大的鮮露筍，細條鮮露筍也可；但口感略遜。

TIPS

• If large thick asparagus is not available, small thin asparagus can be used as well, but the texture is not as good.

全素

木耳紅棗燕麥小米飯
Red Date, Oatmeal and Millet Rice

| 材料 |

木耳1大塊
紅棗4粒
燕麥1量杯*
小米半量杯*
發芽糙米1量杯*
黑蒜3至4粒
*電飯煲量米杯

| 調味料 |

薑黃粉1茶匙
黑椒碎1茶匙
橄欖油1湯匙

| 做法 |

1. 木耳用水浸1小時，洗淨，飛水，過冷河，切條備用。
2. 紅棗去核，洗淨，切細粒。
3. 小米與發芽糙米同洗淨，加入木耳、紅棗、燕麥及適量水，按掣煮飯至熟，再焗5分鐘，加入調味料拌勻，加蓋焗5分鐘即可，上碗後可加入黑蒜同吃。

| INGREDIENTS |

1 large wood ear fungus
4 red dates
1 measuring cup oatmeal*
1/2 measuring cup millet*
1 measuring cup germinated brown rice*
3-4 cloves black garlic
*measuring cup from rice cooker

| SEASONING |

1 tsp turmeric powder
1 tsp finely chopped black pepper
1 tbsp olive oil

| METHOD |

1. Soak wood ear fungus for 1 hour. Rinse, scald and rinse with cold water. Cut into strips.
2. Core red dates, rinse, cut into small dices.
3. Rinse millet and germinated brown rice together in rice cooker. Add wood ear fungus, red dates, oatmeal and water. Cook the rice, remain covered for 5 minutes. Mix in seasoning, remain covered for 5 minutes. Add the black garlic when serving.

煮素小貼士

- 如想吃得香口，最後可拌入麻油進食。

TIPS

- Sesame oil can be added before serving, to give a better aroma.

全素

花生苗煮鹽滷豆腐
Bittern Tofu and Peanut Sprouts

| 材料 |
花生苗半斤
鹽滷豆腐1塊
薑3片（切絲）

| 調味料 |
胡椒粉少許
麻油1茶匙
老抽1茶匙
生抽2茶匙
水半碗

| 獻汁 |
粟粉1茶匙
水2湯匙

| 做法 |
1. 鹽滷豆腐洗淨，切細塊，放入油鍋煎至微黃，盛起。
2. 花生苗洗淨，瀝乾水分。
3. 煎豆腐原鍋下油1湯匙，下薑絲拌香，加入花生苗炒勻，下調味料煮滾，加入鹽滷豆腐煮片刻，最後下獻汁煮滾，上碟食用。

| INGREDIENTS |
300 g peanut sprouts
1 cube bittern tofu
3 slices ginger (shredded)

| SEASONING |
pepper
1 tsp sesame oil
1 tsp dark soy sauce
2 tsp light soy sauce
1/2 bowl water

| THICKENING GLAZE |
1 tsp cornflour
2 tbsp water

| METHOD |
1. Rinse bittern tofu and cut into small pieces. Stir-fry until slightly browned and remove.
2. Rinse peanut sprouts and drain.
3. Add 1 tbsp of oil in the same pan. Add ginger and fry until fragrant. Add peanut sprouts and stir fry. Add seasoning and bring to boil. Put in bittern tofu and cook. Add thickening glaze and bring to boil. Serve.

煮素小貼士

・花生芢質爽味甜，有大量不溶性纖維素，有助腸道蠕動。

TIPS

- Peanut sprouts is crunchy and sweet, its rich insoluble fibre promotes bowel movements.

預防高血壓
Preventing Hypertension

素肉燥冬菇醬番薯湯麵

Sweet Potato Noodles in Vegetarian Minced Pork Sauce

全素

| 材料 |

素肉燥冬菇醬2湯匙（見p.20）

紫番薯幼麵1束

蔬菜少許

昆布素上湯1.5碗（見p.24）

| 做法 |

1. 蔬菜洗淨，備用。
2. 素肉燥冬菇醬蒸5分鐘，保溫。
3. 煮滾昆布素上湯，保溫。
4. 煮滾半鍋水，放入紫番薯麵用中火煮4分鐘，放入蔬菜煮滾，麵及菜放進湯碗，加入素肉燥冬菇醬，淋上昆布素上湯食用。

| INGREDIENTS |

2 tbsp vegetarian minced pork sauce (see p.20)
1 bundle purple sweet potato noodles
vegetable
1.5 bowls kelp stock (see p.24)

| METHOD |

1. Rinse vegetable, set aside.
2. Steam vegetarian minced pork sauce for 5 minutes, keep warm.
3. Bring kelp stock to boil, keep warm.
4. Bring half pot of water to boil. Cook purple sweet potato noodles over medium heat for 4 minutes. Add vegetables and bring to boil. Transfer the noodles and vegetables in a bowl. Add vegetarian minced pork sauce and kelp stock. Serve.

煮素小貼士

- 烹調昆布素上湯的昆布可儲存雪櫃，煮湯麵時加些昆布同吃，非常有益，又不致浪費，尤其預防高血壓方面，建議多吃昆布、海藻類。

TIPS

- Kelp from making kelp stock can be stored in refrigerator; it can be served with the noodles. You don't need to waste the kelp as it is very beneficial. For preventing hypertension, kelp and seaweeds are very effective.

全素

味噌芹菜豆漿豆腐湯
Soymilk, Tofu and Miso Soup

| 材料 |

淡豆漿1.5杯
鹽滷豆腐1塊
芹菜1棵
味噌1湯匙
薑2片
水1.5杯

| 調味料 |

胡椒粉適量
葡萄籽油1茶匙

| 做法 |

1. 芹菜去鬚根，摘去老葉，洗淨，切段。
2. 鹽滷豆腐洗淨，切細塊。
3. 味噌與水2湯匙調勻。
4. 淡豆漿、薑片及水用慢火煮滾，加入味噌煮勻，放入鹽滷豆腐、調味料慢火煮片刻，最後加入芹菜即成。

| INGREDIENTS |

1.5 cups unsweetened soymilk
1 cube bittern tofu
1 stalk Chinese celery
1 tbsp miso
2 slices ginger
1.5 cups water

| SEASONING |

pepper
1 tsp grapeseed oil

| METHOD |

1. Cut off the root from Chinese celery, tear off any old leaves, rinse and cut into sections.
2. Rinse bittern tofu, cut into small pieces.
3. Mix miso and 2 tbsp of water together.
4. Combine soymilk, ginger and water and bring to boil over low heat. Add miso and mix well. Add bittern tofu, seasoning and cook over low heat. Add Chinese celery and serve.

煮素小貼士

・日常多吃鹽滷豆腐，對低密度膽固醇（壞膽固醇）高者有一定幫助，低密度膽固醇減少，血壓自然降低。

TIPS

- Bittern tofu can alleviate high level of low-density lipoprotein cholesterol, which in turn can alleviate hypertension.

全素

蘋果青瓜番茄西芹汁
Apple, Cucumber, Tomato and Celery Juice

| 材料 |

青蘋果1個
青瓜半條
番茄1個
西芹2條

| 做法 |

1. 青蘋果洗淨，去蒂、去籽，切細塊；番茄去蒂，洗淨，切細塊。
2. 青瓜及西芹洗淨，切長條。
3. 青蘋果、番茄、青瓜、西芹分別放進榨汁機，壓成汁即可飲用。

| INGREDIENTS |

1 Granny Smith apple
1/2 cucumber
1 tomato
2 stalks celery

| METHOD |

1. Rinse Granny Smith apple, remove stalk and seeds and cut into small pieces; remove stalk from tomato, rinse and cut into small piece.
2. Rinse cucumber and celery, cut into long strips.
3. Juice Granny Smith apple, tomato, cucumber and celery with a juicer. Serve.

煮素小貼士

・如怕蔬果汁太涼，可加薑2片一同榨汁，以去除寒性。

TIPS

• You can add 2 slices of ginger in the juicer to expel the Cold nature of the juice.

全素

日式七色海藻凍豆腐
Chilled Tofu with Seaweeds in Japanese Style

| 材料 |

日式七色海藻半包

滑豆腐1塊（冷藏）

| 汁料 |（調勻）

麻醬2茶匙

蘋果醋1湯匙

生抽1茶匙

糖半茶匙

| 做法 |

1. 日式七色海藻用凍開水浸10分鐘，瀝乾水分。
2. 滑豆腐放進碗，鋪排七色海藻，淋上汁料，冷藏1小時即可。

煮素小貼士

- 任何海藻類都可配合凍豆腐食用，可依個人喜好而定。
- 常吃海藻與豆腐，有助降低血壓。

TIPS

- All kinds of seaweeds can be used for this dish, use them as desired.
- Having seaweeds and tofu frequently helps lowering blood pressure.

| INGREDIENTS |

1/2 pack Japanese assorted seaweeds
1 soft tofu (chilled)

| SAUCE | (MIXED WELL)

2 tsp sesame paste
1 tbsp apple vinegar
1 tsp light soy sauce
1/2 tsp sugar

| METHOD |

1. Soak assorted seaweeds in cold water for 10 minutes. Drain.
2. Place tofu in a bowl, arrange assorted seaweed on top, and top with the sauce. Refrigerate for 1 hour. Serve.

補鈣、預防骨質疏鬆
Calcium Rich and Preventing Osteoporosis

芝士焗西蘭花
Baked Broccoli with Cheese

蛋奶素

| 材料 |

西蘭花1棵（約12兩）
車打芝士2片（撕細塊）
烘香腰果8粒（壓碎）

| 調味料 |

海鹽半茶匙
油1湯匙

| 做法 |

1. 西蘭花切細朵，放入清水浸半小時，洗淨，瀝乾水分。
2. 西蘭花加入調味料拌勻，隔水大火蒸5分鐘，隔去汁液，鋪上芝士片。
3. 焗爐預熱180℃，放入西蘭花焗10分鐘，待芝士溶化，最後灑下腰果碎即可食用。

| INGREDIENTS |

1 broccoli (about 450 g)
2 slices cheddar cheese (tear into small pieces)
8 roasted cashew nuts (crushed)

| SEASONING |

1/2 tsp sea salt
1 tbsp oil

| METHOD |

1. Cut broccoli into small pieces, soak for 30 minutes, rinse and drain.
2. Mix broccoli with seasoning, steam over high heat for 5 minutes, drain and place cheese on top.
3. Preheat oven at 180°C, bake broccoli for 10 minutes, until cheese dissolves. Top with cashew nuts. Serve.

煮素小貼士

- 蔬菜盡量採用蒸的方法，捨棄用水焓，可保存更多營養成分。
- 可灑上紫菜伴吃，進一步吸收鈣質。

TIPS

- Steaming, instead of boiling, vegetables can preserve more nutrition.
- For absorbing more calcium, top the dish with some seaweeds.

全素

香菇昆布煮鮮枝竹
Simmered Tofu Stick with Black Mushroom and Kelp

| 材料 |

細冬菇8朵
昆布2段（素上湯的昆布）
急凍鮮枝竹4條
薑3片
昆布素上湯1碗（見p.24）

| 調味料 |

素蠔油半湯匙
老抽1茶匙
麻油1茶匙

| 做法 |

1. 冬菇去蒂，用清水浸2小時，洗淨。
2. 急凍鮮枝竹解凍，略洗，瀝乾水分，切段。
3. 昆布切粗條。
4. 燒熱鑊下油1湯匙，下薑片拌香，加入冬菇、昆布素上湯、水1碗，加蓋，用小火煮20分鐘，放入昆布、鮮枝竹煮片刻，下調味料煮勻即成。

| INGREDIENTS |

8 small black mushroom
2 sections kelp (from making kelp stock)
4 frozen tofu stick
3 slices ginger
1 bowl kelp stock (see p.24)

| SEASONING |

1/2 tbsp vegetarian oyster sauce
1 tsp dark soy sauce
1 tsp sesame oil

| METHOD |

1. Remove stalk from black mushrooms, soak for 2 hours, rinse.
2. Defrost tofu stick and rinse. Drain well and cut into section.
3. Cut kelp into thick strips.
4. Heat wok and add 1 tbsp of oil. Stir-fry ginger until fragrant. Add black mushroom, kelp stock, 1 bowl of water. Cover the lid and simmer over low heat for 20 minutes. Add kelp and tofu stick and cook a while. Mix in seasoning. Serve.

煮素小貼士

- 煮完昆布素上湯的昆布不要浪費，以此來煮餸，可口又有益。
- 昆布的鈣質含量豐富，有一定補鈣功效。

TIPS

- Do not waste the kelp after making kelp stock, the kelp from this dish is delicious and beneficial.
- Kelp is rich in calcium and is a great source of it.

全素

馬齒莧菜湯
Purslane Soup

| 材料 |

新鮮馬齒莧菜半斤
板豆腐1塊
蒜肉4粒
大豆芽菜素上湯3碗（見p.23）
海鹽半茶匙

| 做法 |

1. 馬齒莧切去鬚根，洗淨，摘短度。
2. 用少許油炒香蒜肉，加入素上湯煮滾，下豆腐及馬齒莧菜，用中火煮滾片刻，下海鹽調味煮滾，一併喝湯及吃湯料。

| INGREDIENTS |

300 g fresh purslane
1 cube tofu
4 cloves garlic
3 bowls soybean sprout vegetarian stock (see p.23)
1/2 tsp sea salt

| METHOD |

1. Cut off roots from purslane, rinse, tear into short sections.
2. Stir-fry garlic with oil until fragrant. Add soybean sprout vegetarian stock and bring to boil. Add tofu and purslane, bring to boil over medium heat. Season with sea salt and bring to boil. Serve the soup and ingredients.

煮素小貼士

- 馬齒莧菜鈣質高，常吃可補鈣；但含草酸，痛風者慎用。
- 建議購買石膏製成的豆腐，可補充鈣質。

TIPS

- Purslane is rich in calcium; but also contains oxalic acid, people with gout should take caution.
- Tofu made from gypsum have the most calcium.

全素

芝麻黑豆漿
Black Soybean Milk with Sesame

| 材料 |

黑豆半斤
炒香白芝麻1湯匙
冰糖2兩
水9杯

| 做法 |

1. 黑豆洗淨，用水浸過夜；再洗淨黑豆，瀝乾水分。
2. 將黑豆、炒香白芝麻分兩次放入攪拌機，每次加入2杯水，磨成幼滑黑豆漿，過濾（豆渣留用）。
3. 黑豆漿、餘下之水分傾進煲內，打開蓋，用中火煲滾，轉小火，加入冰糖煲15分鐘即可。

煮素小貼士

- 豆渣用慢火烘乾，待冷，放進保鮮袋置雪櫃冷藏，可儲存約10天。
- 當豆漿滾起時，會升高很多，毋須加蓋及轉至小火，以防滿瀉。

TIPS

- Roast the remaining pulps over low heat, let it cool and transfer to zip bag; it can be stored in refrigerator up to 10 days.
- Use low heat to cook the soymilk without covering the pot, as the soymilk is easy to boil over.

| INGREDIENTS |

300 g black soybean
1 tbsp toasted sesame
75 g rock sugar
9 cups water

| METHOD |

1. Rinse black soybeans and soak overnight. Rinse and drain.
2. Put black soybeans and sesame in a blender in two batches, add 2 cups of water for each batch, and blend into smooth milk. Strain and keep the pulps.
3. Put the milk and remaining water in a pot. Bring to boil over medium heat without covering the pot. Turn to low heat, add rock sugar and simmer for 15 minutes. Serve.

蛋奶素

豆渣甘筍菠菜煎餅
Fried Egg with Spinach, Carrot and Soy Pulp

| 材料 |

炒乾豆渣1碗

菠菜4兩

甘筍半條

雞蛋1隻（拂勻）

粟粉1湯匙

| 調味料 |

胡椒粉少許

海鹽半茶匙

| 做法 |

1. 菠菜用水浸半小時，切掉鬚根，洗淨，飛水，過冷河，擠乾水分，切碎。
2. 甘筍去皮，洗淨，刨絲。
3. 乾豆渣、菠菜、甘筍、蛋漿、調味料，粟粉拌勻。
4. 燒熱鑊下油3湯匙，舀1湯匙豆渣餅漿進鍋內，煎至兩面金黃香脆即可。

| INGREDIENTS |

1 bowl roasted soy pulp
150 g spinach
1/2 carrot
1 egg (whisked)
1 tbsp cornflour

| SEASONING |

pepper
1/2 tsp sea salt

| METHOD |

1. Soak spinach for 30 minutes and cut off the roots. Rinse and scald. Rinse with cold water, squeeze until dry and chop.
2. Peel carrot, rinse and shred.
3. Mix roasted soy pulp, spinach, carrot, egg, seasoning and cornflour together.
4. Heat wok and add 3 tbsp of oil, fry 1 tbsp of the egg mixture until both side browned. Serve.

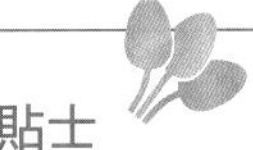

煮素小貼士

- 豆渣可補鈣，用來煎餅，美味又可口。烘乾的豆渣可放冰箱儲存。

TIPS

- Soy pulp is a great source of calcium and is delicious for making fried egg. Roasted soy pulp can be stored in refrigerator.

補腦、提升記憶力
Strengthening Brain and Memory

薑黃合桃芝麻糊
Walnut and Sesame Sweet Soup

全素

材料

烘香合桃2兩
烘香黑芝麻3兩
薑黃粉1茶匙
粘米2湯匙
冰糖2兩
水3杯

做法

1. 合桃、黑芝麻、粘米放入攪拌機，加水3杯，磨成幼滑的黑芝麻漿，過濾。
2. 合桃黑芝麻漿用中小火煮滾，加入冰糖，用小火煮約10分鐘，熄火，加入薑黃粉拌勻即可。

INGREDIENTS

75 g toasted walnuts
113 g toasted black sesame
1 tsp turmeric powder
2 tbsp white rice
75 g rock sugar
3 cups water

METHOD

1. Blend walnuts, black sesame, white rice, 3 cups of water in a blender into a smooth sesame mixture. Strain.
2. Bring the mixture to boil over medium low heat, add rock sugar and cook over low heat for 10 minutes. Turn off heat. Add turmeric powder and mix well. Serve.

煮素小貼士

- 薑黃粉簡單食用法：用1/3茶匙薑黃粉沖入1杯溫水，加1茶匙蜜糖拌勻飲用。
- 隔天飲用1杯薑黃水，有助提升記憶力。

TIPS

- Simple drink with turmeric powder: mix 1/3 tsp turmeric powder, 1 cup of warm water, 1 tsp of honey together and serve.
- Having a cup of turmeric drink helps improving memory.

全素

菠蘿合桃甘筍炒素雞

Stir-fried Tofu Skin Roll with Pineapple, Walnut and Carrot

| 材料 |

大素雞1條
新鮮菠蘿半個
甘筍半個
烘香合桃肉3湯匙
乾葱2粒（去衣、切片）

汁料

番茄汁3湯匙
檸檬汁1湯匙
黃砂糖1茶匙
鹽半茶匙
水2湯匙

| 做法 |

1. 甘筍去皮，切片，放入滾水灼2分鐘，撈起備用。
2. 素雞洗淨，切滾刀塊；菠蘿切細塊。
3. 燒熱鑊下油2湯匙，下素雞煎至金黃，加入乾葱炒香，下菠蘿、甘筍、汁料煮滾拌勻，加入獻汁炒勻煮滾即可，上碟，灑入合桃食用。

| INGREDIENTS |

1 large tofu skin roll (vegetarian chicken)
1/2 fresh pineapple
1/2 carrot
3 tbsp toasted walnut
2 shallots (peeled, sliced)

| SAUCE |

3 tbsp ketchup
1 tbsp lemon juice
1 tsp brown sugar
1/2 tsp salt
2 tbsp water

| METHOD |

1. Peel and slice carrot. Scald for 2 minutes and set aside.
2. Rinse tofu skin roll, cut at an angle into pieces; cut pineapple into small pieces.
3. Heat wok and add 2 tbsp of oil. Stir-fry tofu skin roll until golden brown. Stir-fry shallots until fragrant. Add pineapple, carrot and sauce, bring to boil and mix well. Add thickening glaze, bring to boil. Plate and top with walnut. Serve.

菠蘿合桃甘筍炒素雞

煮素小貼士

・合桃補腦，加入日常餸菜中，美味又有益身體。

TIPS

- Walnuts improve brain function, it is beneficial and delicious to add walnuts in everyday dishes.

全素

薑黃松子仁意式三色藜麥飯
Quinoa with Turmeric, Pine Nuts in Italian Style

| 材料 |

三色藜麥1量杯*
松子仁2湯匙
薑黃粉1茶匙
青瓜片適量
番茄1個（切碎）
甘筍半個（切碎）
洋葱1/4個（切碎）
*電飯煲量米杯

| 調味料 |

乾羅勒碎2茶匙
黑胡椒碎半茶匙
海鹽1/3茶匙

| 做法 |

1. 下橄欖油2湯匙，放入洋葱、番茄、甘筍炒香，加入三色藜麥、調味料、清水1碗煮滾，加蓋小火煮15分鐘至水乾，熄火，焗10分鐘，下薑黃粉拌勻。
2. 盛起藜麥飯，配上松仔仁、青瓜片食用。

| INGREDIENTS |

1 measuring cup three-color quinoa*
2 tbsp pine nuts
1 tsp turmeric powder
cucumber slices
1 tomato (chopped)
1/2 carrot (chopped)
1/4 onion (chopped)
*measuring cup from rice cooker

| SEASONING |

2 tsp chopped dried basil
1/2 tsp finely chopped black pepper
1/3 tsp sea salt

| METHOD |

1. Stir-fry onion, tomato, carrot in a pan with olive oil until fragrant. Add quinoa, seasoning, 1 bowl of water and bring to boil. Cover the lid and cook over low heat for 15 minutes, until dry. Turn off heat, remain covered for 10 minutes. Mix in turmeric powder.
2. Mix cooked quinoa with pine nuts and cucumber slices. Serve.

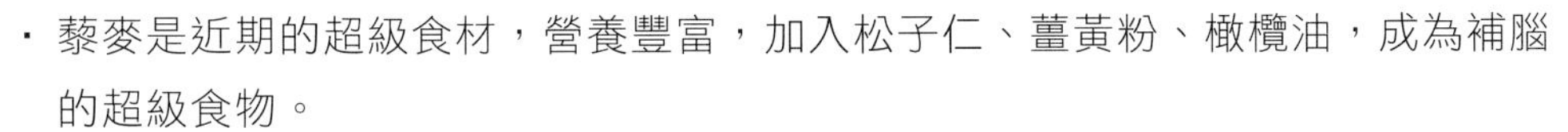

煮素小貼士

- 藜麥是近期的超級食材，營養豐富，加入松子仁、薑黃粉、橄欖油，成為補腦的超級食物。

TIPS

- Quinoa is a recent superfood; it is very nutritious. Dishes with quinoa, pine nuts, olive oil are perfect for strengthening brain.

全素

麻油拌炒銀杏豆乾毛豆仁

Stir-fried Edamame, Dried Tofu, Ginkgo with Sesame Oil

| 材料 |

銀杏2兩
毛豆仁4兩
粟粒半碗
豆乾2塊
鮮百合1球
乾葱2粒（去衣、切片）

| 調味料 |

胡椒粉少許
黃砂糖半茶匙
生抽半湯匙

| 做法 |

1. 銀杏飛水，過冷河，去外衣。
2. 毛豆仁飛水3分鐘，撈起，瀝乾水分。
3. 豆乾洗淨，切粒。
4. 鮮百合撕開一瓣瓣，浸於水中洗淨，撈起，瀝乾水分。
5. 燒熱鑊下麻油1湯匙，加入乾葱拌香，放下銀杏、毛豆仁、粟粒、豆乾炒勻，加入熱水3湯匙煮片刻，下調味料、鮮百合炒勻，最後下麻油2茶匙拌勻，上碟享用。

| INGREDIENTS |

75 g ginkgo
150 g edamame
1/2 bowl corn kernels
2 pieces dried tofu
1 fresh lily bulb
2 shallots (peeled, sliced)

| SEASONING |

pepper
1/2 tsp brown sugar
1/2 tbsp light soy sauce

| METHOD |

1. Scald ginkgo, rinse with cold water and peel.
2. Scald edamame for 3 minutes, remove and drain.
3. Rinse and dice dried tofu.
4. Tear lily bulbs into pieces, soak and rinse. Drain.
5. Heat wok and add 1 tbsp of sesame oil. Stir-fry shallots until fragrant. Add ginkgo, edamame, corn kernels and dried tofu and stir well. Add 3 tbsp of hot water and cook. Mix in seasoning and lily bulbs. Add 2 tsp of sesame oil and mix well. Serve.

煮素小貼士

- 無論是銀杏、毛豆仁最宜先焓後煮，煮至全熟才進食。銀杏、毛豆仁營養豐富，經常適量進食有助提升記憶力。

TIPS

- Ginkgo and edamame are best scalded and cooked before use. Both ginkgo and edamame are nutritious, they help improving memory functions.

全素

雪菜鮮冬菇芽菜拌薑黃麵

Turmeric Noodles with Salted Mustard Green, Black Mushroom and Mung Bean Sprout

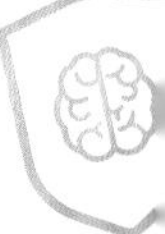

| 材料 |

薑黃麵1束

雪菜1小棵

鮮冬菇4朵

芽菜4兩

薑2片（切絲）

| 汁料 |

麻油2茶匙

素蠔油半湯匙

| 做法 |

1. 煮滾清水半鍋，放入薑黃麵拌散，焓7分鐘，瀝乾水分。
2. 雪菜用水浸15分鐘，擠乾水分，切碎；鮮冬菇去蒂，洗淨，切厚片；芽菜洗淨，隔乾水分。
3. 燒熱鑊下油1湯匙，下薑絲、雪菜炒香，加入冬菇、芽菜炒片刻，盛起。
4. 薑黃麵加入汁料拌勻，鋪上雪菜鮮冬菇、芽菜食用。

| INGREDIENTS |

1 bundle turmeric noodle

1 small stalk salted mustard green

4 fresh black mushrooms

150 g mung bean sprouts

2 slices ginger (shredded)

| SAUCE |

2 tsp sesame oil

1/2 tbsp vegetarian oyster sauce

| METHOD |

1. Bring half pot of water to boil. Spread the turmeric noodle and boil for 7 minutes. Drain.
2. Soak salted mustard green for 15 minutes. Squeeze until dry and chop; remove stalks from black mushrooms, rinse and cut into thick slices; rinse mung bean sprouts and drain.
3. Heat wok and add 1 tbsp of oil. Stir-fry ginger, salted mustard green until fragrant. Add black mushrooms and mung bean sprout and stir fry. Remove.
4. Mix turmeric noodles with the sauce. Top with salted mustard green, black mushrooms and mung bean sprouts. Serve.

煮素小貼士

- 薑黃食用多元化，印度咖喱含薑黃成分多，可炒薑黃飯、薑黃拌飯、拌麵等。
- 薑黃麵含有薑黃元素，是另類之補腦麵食。
- 如喜歡薑黃的味道，享用前可灑入少許薑黃粉。

TIPS

- Turmeric is versatile and can be used in various dishes. Turmeric is one of the main ingredients for Indian curry. Turmeric can be used to cook rice or serve with rice and noodles.
- Turmeric noodles contain curcumin, is an alternative choice for strengthening brain.
- If you like the taste of turmeric, you can mix in some turmeric powder before serving.

養肝明目
Tonifying Liver, Improving Eyesight

煮素小貼士

- 多進食橙色及黃色的食材，如：南瓜、甜椒、甘筍、番茄，對眼睛有黃斑點病變的人士有改善作用。

TIPS

- Orange and yellow ingredients, such as pumpkin, bell pepper, carrot, tomato, can alleviate macular degeneration.

全素

鮮蔬燉南瓜
Stewed Pumpkin with Assorted Vegetables

| 材料 |

南瓜半斤
西蘭花4兩
四季豆4兩
甜椒半個
番茄1個
洋葱半個
蒜肉2粒（切片）

| 調味料 |

海鹽半茶匙
紅椒粉1茶匙

| 做法 |

1. 南瓜去籽，洗淨，切大塊。
2. 西蘭花切細朵，用水浸半小時，洗淨。
3. 四季豆撕去老筋，洗淨，切段。
4. 甜椒去籽，洗淨，切塊；番茄去蒂，洗淨，切角；洋葱去衣，洗淨，切碎。
5. 燒熱鑊下油2湯匙，下蒜肉、洋葱拌香，加入南瓜、水3碗，用中火煮15分鐘，下番茄、四季豆、西蘭花、甜椒用小火再煮8分鐘，拌入調味料煮滾即成。

| INGREDIENTS |

300 g pumpkin
150 g broccoli
150 g snap beans
1/2 bell pepper
1 tomato
1/2 onion
2 cloves garlic (sliced)

| SEASONING |

1/2 tsp sea salt
1 tsp paprika

| METHOD |

1. Remove seeds from pumpkin, rinse and cut into large pieces.
2. Cut broccoli into small pieces. Soak for 30 minutes and rinse.
3. Tear off hard vein from snap beans. Rinse and cut into sections.
4. Remove seeds from bell pepper, rinse and cut into pieces; remove stalk from tomato, rinse, cut in quarters; peel onion, rinse and chop.
5. Heat wok and add 2 tbsp of oil. Stir-fry garlic, onion until fragrant. Add pumpkin, 3 bowls of water and cook over medium heat for 15 minutes. Add tomato, snap beans, broccoli, bell pepper and cook over low heat for 8 minutes. Mix in seasoning and bring to boil. Serve.

全素

杞子枸杞菜湯

Qi Zi and Matrimony Vine Soup

| 材料 |

枸杞菜半斤

杞子1湯匙

大豆芽菜素上湯3碗（見p.23）

| 調味料 |

海鹽半茶匙

油1茶匙

| 做法 |

1. 枸杞菜摘取菜葉，洗淨，瀝乾水分。
2. 杞子用清水沖洗，瀝乾水分。
3. 煮滾大豆芽菜素上湯，下調味料，加入枸杞菜煮滾片刻，最後下杞子煮滾即成。

| INGREDIENTS |

300 g matrimony vine
1 tbsp Qi Zi
3 bowl soybean sprout vegetarian stock (see p.23)

| SEASONING |

1/2 tsp sea salt
1 tsp oil

| METHOD |

1. Take the leaves from matrimony vine, rinse and drain.
2. Rinse Qi Zi and drain.
3. Bring soybean sprout vegetarian stock to boil. Add seasoning and matrimony leaves and bring to boil. Add Qi Zi and bring to boil. Serve.

煮素小貼士

・本湯水清肝明目，尤其每天長時間對着電腦工作，或使用手機人士，常飲有益。

・緊記杞子最後才加入，以免湯水帶酸味。

TIPS

- This soup tonifies liver and improve eyesight, it is particularly beneficial for people use computer and mobile phone for a long time.
- To avoid the soup turning sour, remember to add Qi Zi last.

杞子枸杞菜湯

全素

香烤小甘筍南瓜件
Roasted Baby Carrot and Pumpkin in Pesto Sauce

| 材料 |

小甘筍4條

南瓜半斤

欖油羅勒醬適量（見p.16）

| 做法 |

1. 小甘筍外皮洗淨；南瓜洗淨，連皮切塊。
2. 將小甘筍、南瓜件排上焗盤，掃上一層橄欖油，放入已預熱之焗爐，用160°C焗30分鐘，取出，放置10分鐘，配以欖油羅勒醬供食。

煮素小貼士

・小甘筍含胡蘿蔔素，進入身體後轉化成維他命A，是眼睛的營養品，對黃斑點病變者有一定助益。

・也可烤焗橙黃甜椒，香甜味美，而且維他命A豐富。

TIPS

- Carotene in baby carrot will transform into vitamin A, which is nutrition for our eyes. It has a certain alleviation on macular degeneration.
- You can bake orange and yellow bell pepper too, they are sweet taste and full of vitamin A.

| INGREDIENTS |

4 baby carrots
300 g pumpkin
pesto sauce (see p.16)

| METHOD |

1. Rinse baby carrots; rinse pumpkin and cut into pieces.
2. Arrange baby carrots, pumpkin on a baking tray. Spread with olive oil. Bake in a preheated oven at 160°C for 30 minutes. Remove, set aside for 10 minutes. Serve with pesto sauce.

全素

藍莓木瓜乾果伴合桃火箭菜

Rocket Salad with Blueberry, Papaya, Dried Berries and Walnut

| 材料 |

新鮮藍莓15粒

熟木瓜半個

火箭菜適量

烘香合桃肉1湯匙

紅莓乾、提子乾共1湯匙

| 汁料 |（調勻）

檸檬汁半湯匙

橄欖油1湯匙

海鹽少許

| 做法 |

1. 火箭菜洗淨，瀝乾水分。
2. 藍莓洗淨，抹乾水分；木瓜去皮、去籽，切細塊。
3. 火箭菜鋪上碟，排上木瓜、藍莓，淋上汁料，伴紅莓乾、提子乾及合桃食用。

| INGREDIENTS |

15 fresh blueberries

1/2 mature papaya

rocket salad

1 tbsp toasted walnuts

1 tbsp dried cranberries & raisin

| SAUCE | (MIXED WELL)

1/2 tbsp lemon juice

1 tbsp olive oil

sea salt

| METHOD |

1. Rinse rocket salad and drain.
2. Rinse blueberries, wipe until dry; skin papaya and remove seeds, cut into small pieces.
3. Arrange rocket salad on a plate, put on papaya and blueberries. Top with the sauce and serve with dried cranberries, raisin and walnut.

煮素小貼士

- 藍莓含很豐富的花青素，有助恢復眼睛疲勞，對白內障、眼澀及眼睛模糊有幫助。
- 可配搭任何乾果及果仁；但果仁建議先烘香，香脆可口。

TIPS

- Blueberries are rich in anthocyanin; it can alleviate eye exhaustion, cataract, sore eyes and blurry vision.
- Any dried fruits and nuts can be used; it is suggested to toast nuts first to make it crunchy and delicious.

改善痛風
Relieving Gout

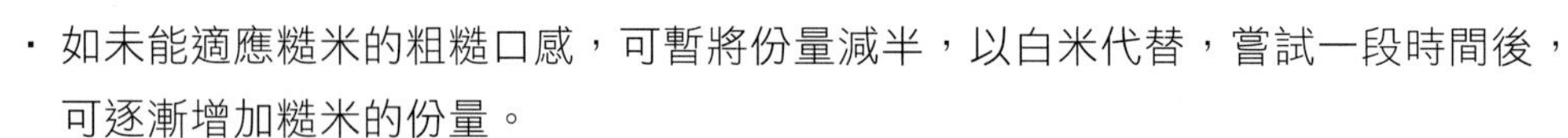

煮素小貼士

・如未能適應糙米的粗糙口感，可暫將份量減半，以白米代替，嘗試一段時間後，可逐漸增加糙米的份量。

TIPS

- If you are used to the rough texture of brown rice, replace half with white rice. Add the amount brown rice gradually after a certain amount of time.

全素

腰果青瓜紅藜糙米飯
Brown Rice, Quinoa with Cashew and Cucumber

| 材料 |

紅藜麥1/3量杯*
糙米1.5量杯*
烘香腰果3湯匙
甘筍半條
小青瓜1條
*電飯煲量米杯

| 調味料 |

海鹽1/4茶匙
橄欖油1湯匙

| 做法 |

1. 糙米洗淨，用清水浸2小時。
2. 甘筍去皮，洗淨，切粒；青瓜切去頭尾兩端，洗淨，切粒。
3. 將糙米、紅藜麥、甘筍放進電飯煲，加入適量水，按掣煮飯至飯熟透，再焗10分鐘，下調味料、青瓜拌勻，盛起米飯，伴腰果食用。

| INGREDIENTS |

1/3 measuring cup red quinoa*
1.5 measuring cups brown rice*
3 tbsp toasted cashew nuts
1/2 carrot
1 small cucumber
*measuring cup from rice cooker

| SEASONING |

1/4 tsp sea salt
1 tbsp olive oil

| METHOD |

1. Rinse brown rice and soak for 2 hours.
2. Peel carrot, rinse and dice; cut off both ends from cucumber, rinse and dice.
3. Put brown rice, red quinoa, carrot in rice cooker. Add water and cook the ingredients until fully cooked. Remain covered for 10 minutes. Mix in seasoning and cucumber. Remove the rice from rice cooker and serve with cashew nuts.

蛋奶素

欖油蘋果醋伴烚蔬菜
Poached Vegetables with Olive Oil and Apple Vinegar

| 材料 |

椰菜1/4個

馬鈴薯2個

甘筍1條

雞蛋2隻

黑橄欖4粒

蒜茸1湯匙

| 調味料 |

橄欖油1湯匙

蘋果醋1湯匙

| 做法 |

1. 馬鈴薯、甘筍去皮，洗淨。
2. 馬鈴薯、雞蛋放入煲，注入蓋過馬鈴薯約3cm之清水，用中火煲滾，再煲10分鐘，撈起雞蛋，放入甘筍、椰菜煲15分鐘，隔去水分。
3. 雞蛋放入清水浸凍，去殼，切成兩邊。
4. 馬鈴薯、甘筍及椰菜切塊。
5. 馬鈴薯、甘筍、椰菜、雞蛋、黑橄欖上碟，灑入蒜茸，淋上調味料即可。

| INGREDIENTS |

1/4 cabbage
2 potatoes
1 carrot
2 eggs
4 black olives
1 tbsp grated garlic

| SEASONING |

1 tbsp olive oil
1 tbsp apple vinegar

| METHOD |

1. Peel potatoes and carrot, rinse.
2. Put potatoes and eggs in a pot. Pour in water about 3cm higher than potatoes. Bring to boil over medium heat, boil for 10 minutes. Remove eggs. Add carrot, cabbage and boil for 15 minutes. Drain.
3. Soak eggs until cold, shell and cut in half.
4. Cut potatoes, carrot and cabbage into pieces.
5. Arrange potatoes, carrot, cabbage, eggs and black olives on a plate. Top with grated garlic and seasoning. Serve.

煮素小貼士

- 焓蔬菜配以橄欖油、蘋果醋、蒜茸同吃，開胃、美味又有益健康；但有胃酸者慎吃。
- 焓蔬菜時下點鹽粒，增添焓菜的味道。

TIPS

- It is appetising, delicious and healthy to serve poached vegetables with olive oil, apple vinegar and grated garlic. People with acid reflux should take cautions.
- Poach vegetables with a little salt can strengthen their taste.

全素

木耳甘筍炒萵筍絲
Stir-fried Celtuce, Wood Ear Fungus and Carrot

| 材料 |

木耳1大塊
甘筍半條
萵筍1條（約12兩）
炒香白芝麻2茶匙
薑2片（切絲）

| 調味料 |

黃砂糖半茶匙
生抽2茶匙
麻油1茶匙

| 做法 |

1. 木耳用水浸2小時，飛水，過冷河，瀝乾水分。
2. 萵筍削去硬皮，洗淨；甘筍去皮，洗淨。
3. 木耳、甘筍、萵筍分別切絲。
4. 燒熱鑊下油1湯匙，下薑絲抨香，加入木耳、甘筍炒片刻，放入萵筍、熱水2湯匙炒片刻，下調味料炒勻，上碟，最後灑上白芝麻即成。

| INGREDIENTS |

1 large wood ear fungus
1/2 carrot
1 celtuce (about 450 g)
2 tsp toasted sesame
2 slices ginger (shredded)

| SEASONING |

1/2 tsp brown sugar
2 tsp light soy sauce
1 tsp sesame oil

| METHOD |

1. Soak wood ear fungus for 2 hours. Scald, rinse with cold water and drain.
2. Scrape off hard skin from celtuce, rinse; peel of carrot, rinse.
3. Shred wood ear fungus, carrot, celtuce.
4. Heat wok and 1 tbsp of oil. Stir-fry ginger until fragrant. Add wood ear fungus and carrot and stir-fry. Add celtuce and 2 tbsps of hot water and stir-fry. Add seasoning and stir well. Plate and top with sesame. Serve.

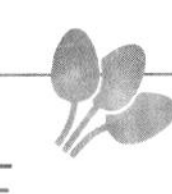

煮素小貼士

- 萵筍比較硬實，削皮切絲時需要留心，以免弄傷。
- 這是一款清爽美味的菜式，可配糙米飯、米粉、麵等進食，美味又健康。

TIPS

- Celtuce is relatively hard. Careful when scraping and shredding to avoid injury.
- This is a tasty, refreshing dish. It can be served with brown rice, rice vermicelli or noodles; it is healthy and delicious.

全素

欖油檸汁雅枝竹
Artichoke with Olive Oil and Lemon Juice

| 材料 |

新鮮雅枝竹1個

麥包3片

| 汁料 | （調勻）

海鹽1/8茶匙

檸檬汁1湯匙

橄欖油2湯匙

| 做法 |

1. 雅枝竹撕去外葉，留嫩葉及芯，原個放在滾水，用小火烚半小時，撈起。
2. 雅枝竹待暖，切片，加入汁料拌勻，備用。
3. 麥包放入焗爐烘脆，雅枝竹排在麥包上伴吃。

| INGREDIENTS |

1 fresh artichoke

3 slices whole wheat bread

| SAUCE | (MIXED WELL)

1/8 tsp sea salt

1 tbsp lemon juice

2 tbsp olive oil

| METHOD |

1. Tear off outer leaves, keep the baby leaves and the core as a whole. Boil the artichoke over low heat for 30 minutes, remove.
2. Let the artichoke warm, slice and mix well with the sauce. Set aside.
3. Bake the whole wheat bread until crispy. Arrange artichoke on top of the bread. Serve.

煮素小貼士

- 雅枝竹的硬外葉，與雅枝竹芯一同煲熟，待冷，切碎，再放入攪拌機，加入適量橄欖油、檸檬汁、少許海鹽磨成糊，可伴全麥包進食。

TIPS

- The hard outer leaves of the artichoke can be boiled together. Let them cool and chop. Blend the artichoke leaves, olive oil, lemon juice and sea salt together, serve with whole wheat bread.

改善糖尿
Relieving Diabetes

全素

松子仁拌洋葱紅椰菜
Stir-fried Red Cabbage with Pine Nuts and Onion

| 材料 |

紅椰菜6兩
洋葱半個
松子仁2湯匙
蒜肉2粒（拍碎）

| 調味料 |

黑椒碎半湯匙
海鹽1/4茶匙

| 做法 |

1. 紅椰菜撕成塊，洗淨，切絲；洋葱去衣，洗淨，切絲。
2. 平底鑊下橄欖油2湯匙，下蒜肉、洋葱炒香，加入紅椰菜炒勻，下熱水3湯匙煮片刻，加入調味料拌勻，上碟，灑下松子仁即可。

| INGREDIENTS |

225 g red cabbage
1/2 onion
2 tbsp pine nuts
2 cloves garlic (crushed)

| SEASONING |

1/2 tbsp black pepper
1/4 tsp sea salt

| METHOD |

1. Tear red cabbage into pieces, rinse and shred; peel onion, rinse and shred.
2. Heat wok and add 2 tbsp of olive oil. Stir-fry garlic, onion until fragrant. Add red cabbage and stir-fry. Add 3 tbsp of hot water and cook. Mix in seasoning, plate and top with pine nuts. Serve.

煮素小貼士

- 糖尿病者應多進食各種類的蔬菜，如椰菜、西蘭花、菜心、生菜，熟菜和生菜都可吃。

TIPS

- People with diabetes should have various vegetables, like cabbage, broccoli, choy sum, lettuce, both raw and cooked.

昆布素上湯蕎麥麵
Buckwheat Noodles in Kelp Stock

| 材料 |

蕎麥麵1束

乾燥海藻1湯匙

菜心4條

昆布素上湯1.5碗（見p.24）

海鹽半茶匙

| 調味料 |

生抽1茶匙

麻油1茶匙

| 做法 |

1. 菜心洗淨，瀝乾水分。
2. 乾燥海藻用凍開水浸10分鐘，瀝乾水分。
3. 煮滾清水半鍋，放進蕎麥麵、海鹽半茶匙焓6分鐘，撈起蕎麥麵，瀝乾水分。
4. 煮滾昆布素上湯，放入菜心滾片刻即可；蕎麥麵放入湯碗，傾下昆布素上湯、菜心、海藻、調味料即可。

| INGREDIENTS |

1 bundles buckwheat noodles

1 tbsp dried seaweed

4 stalks choy sum

1.5 bowls kelp stock (see p.24)

1/2 tsp sea salt

| SEASONING |

1 tsp light soy sauce

1 tsp sesame oil

| METHOD |

1. Rinse choy sum and drain.
2. Soak dried seaweeds in cold water for 10 minutes. Drain.
3. Heat half pot of water, add 1/2 tsp sea salt and boil buckwheat noodles for 6 minutes. Remove and drain.
4. Bring kelp stock to boil, scald choy sum; transfer buckwheat noodles in a soup bowl, add kelp stock, choy sum, dried seaweed and seasoning. Serve.

煮素小貼士

- 可煮蕎麥麵、菠菜麵，配合海藻和蔬菜進食；但糖尿病者要注意麵的份量，每餐不宜進食太多澱粉質，宜少食多餐。

TIPS

- Buckwheat and spinach noodles can be served with seaweeds and vegetable; be caution about the amount of noodles, the diabetes patient avoids having too much starch in a meal. It is suggested to have frequent and small portion of meal.

全素

鮮蔬拌豆茸餅

Bean Patties with Vegetables

| 材料 |

素豆餅3塊（見p.29）
沙律菜適量

| 汁料 |（調勻）

海鹽1/8茶匙
檸檬汁半湯匙
橄欖油1茶匙

| 做法 |

1. 素豆餅放平底鍋，用慢火烘至兩面金黃，備用。
2. 沙律菜洗淨，瀝乾水分，上碟。
3. 放上素豆餅，淋上汁料食用。

| INGREDIENTS |

3 pieces bean patties (see p.29)
salad vegetables

| SAUCE | (MIXED WELL)

1/8 tsp sea salt
1/2 tbsp lemon juice
1 tsp olive oil

| METHOD |

1. Fry bean patties over low heat until both sides turn gold brown. Set aside.
2. Rinse vegetables, drain and plate.
3. Top with bean patties and the sauce. Serve.

煮素小貼士

- 素豆餅營養豐富，捏塊狀後宜冷藏至略硬，減低煎煮時弄散。
- 素豆餅配合適量沙律菜，可作為午餐或晚餐食用。

TIPS

- Bean patties are nutritious. After shaping, they should be refrigerated until slightly hardened; they will keep their shapes better when frying.
- This dish can serve as lunch or dinner.

粟粒菜梗豆乾糙米飯

Brown Rice with Dried Tofu, Corn and Broccoli Stalk

| 材料 |

粟粒半碗

西蘭花菜梗2塊

豆乾2塊

糙米2量杯*

乾葱2粒（去衣、切片）

*電飯煲量米杯

| 調味料 |

生抽2茶匙

橄欖油1湯匙

| 做法 |

1. 糙米洗淨，用水浸2小時。
2. 西蘭花菜梗撕去外層，洗淨，切粗粒。
3. 豆乾洗淨，切粗粒。
4. 燒熱鑊下油1湯匙，下乾葱炒香，加入西蘭花梗、粟粒、豆乾炒片刻，盛起備用。
5. 糙米放進電飯煲，加入適量清水，按掣煮至飯大滾，加入粟粒、菜梗、豆乾，煮至飯熟再焗10分鐘，加入調味料拌勻即可。

| INGREDIENTS |

1/2 bowl corn kernels

2 pieces broccoli stalks

2 pieces dried tofu

2 measuring cups brown rice*

2 shallots (peeled, sliced)

*measuring cup from rice cooker

| SEASONING |

2 tsp light soy sauce

1 tbsp olive oil

| METHOD |

1. Rinse brown rice and soak for 2 hours.
2. Tear off hard skin from broccoli stalks, rinse and cut into large dice.
3. Rinse dried tofu and cut into large dice.
4. Heat wok and add 1 tbsp of oil. Stir-fry shallots until fragrant. Add broccoli stalks, corn kernels and dried tofu and stir-fry. Set aside.
5. Put brown rice in rice cooker, add water and cook brown rice. When the rice boils, add corn kernels, broccoli stalks, dried tofu. When the rice is fully cooked, remain covered for 10 minutes. Mix in seasoning and serve.

煮素小貼士

・糖米飯升糖指數低，常吃有益；如嫌糙米飯粗糙，可選用發芽糙米，飯質軟滑點，也是一種好選擇。

TIPS

- Brown rice has a low glycemic index, it is great to have it frequently; if brown rice is too rough, you can also use germinated brown rice, which is softer.

100道素菜 *Vegetarian Meals*

作者 Author
杜紹鵬 To Siu Pang

策劃/編輯 Project Editor
簡詠怡 Karen Kan

攝影 Photographer
細權 Leung Sai Kuen

美術設計 Design
羅穎思 Venus Lo

出版者 Publisher
Forms Kitchen
香港鰂魚涌英皇道1065號 東達中心1305室
Room 1305, Eastern Centre, 1065 King's Road, Quarry Bay, Hong Kong.
電話 Tel: 2564 7511
傳真 Fax: 2565 5539
電郵 Email: info@wanlibk.com
網址 Web Site: http://www.wanlibk.com
http://www.facebook.com/wanlibk

發行者 Distributor
香港聯合書刊物流有限公司 SUP Publishing Logistics (HK) Ltd.
香港新界大埔汀麗路36號 中華商務印刷大廈3字樓
3/F., C&C Building, 36 Ting Lai Road, Tai Po, N.T., Hong Kong
電話 Tel: 2150 2100
傳真 Fax: 2407 3062
電郵 Email: info@suplogistics.com.hk

承印者 Printer
中華商務彩色印刷有限公司 C&C Offset Printing Co., Ltd.

出版日期 Publishing Date
二〇一九年七月第一次印刷 First print in July 2019

Published in Hong Kong by Forms Kitchen,
a division of Wan Li Book Company Limited.
ISBN 978-962-14-7046-1